Number Theory - Structures, Examples and Problems

Number Theory - Structures, Examples and Problems

Editor

Nik Pachis

scitus
academics

Number Theory - Structures, Examples and Problems
Edited by **Nik Pachis**

ISBN: 978-1-68117-447-1
Library of Congress Control Number: 2016937604

SCITUS Academics LLC,
www.scitusacademics.com
Box No. 4766, 616 Corporate Way,
Suite 2, Valley Cottage,
NY 10989

Printed in United States of America on Acid Free Paper

Preface

Number theory, branch of mathematics concerned with properties of the positive integers. Sometimes called "higher arithmetic," it is among the oldest and most natural of mathematical pursuits. Number theory has always fascinated amateurs as well as professional mathematicians. In contrast to other branches of mathematics, many of the problems and theorems of number theory can be understood by laypersons, although solutions to the problems and proofs of the theorems often require a sophisticated mathematical background. Number theory is a vast and fascinating field of mathematics, sometimes called "higher arithmetic," consisting of the study of the properties of whole numbers. Primes and prime factorization are especially important in number theory, as are a number of functions such as the divisor function, Riemann zeta function, and totient function. Number theory is an ongoing rich area of mathematical exploration with connections and applications to other fields from representation theory, to physics, cryptography, and more. While the forefront of number theory is replete with sophisticated and famous open problems, at its foundation are basic, elementary ideas that can stimulate and challenge beginning students. This book Number Theory - Structures, Examples and Problems focuses on a problem-solving approach to the subject.

Table of Contents

CHAPTER 1

Development of New Method for Generating Prime Numbers

Seidikassym Baibekov[1], Serik Altynbek[2]

[1]Research Institute of Coal Chemistry and Technology, Astana, Kazakhstan

[2]Eurasian National University, Astana, Kazakhstan

ABSTRACT

The article is devoted to actual problems of prime numbers. A theorem that allows generating a sequence of prime numbers is proposed. An algorithm for generating prime numbers has been developed. A comparison of the proposed theorem, with Wilson's theorem is also provided.

KEYWORDS

Prime Numbers, Theorem, Algorithm, Method, Prime Twins, Generation

1. INTRODUCTION

The reason for writing this article was a solution of the ancient problem. This problem in a simplified version is as follows: Slandy a noble woman well-known in the Eastern world lived in ancient times. She had seven daughters. Slandy always wore aтamazing beauty antique necklace of precious pearls, which according to tradition passed from mother-in-law to daughter-in-law. In old age, she told her daughters-in-law: "By inheritance it is time to pass the necklace to someone of you and if I will choose someone of you, the others will be offended. If I choose two of you and divide this necklace exactly into two parts, one pearl will be surplus. This is not a right way plus other fives will feel aggrieved what I don't want in my old ages. And also, when this necklace is divided into 3, 4 or 5, and 6, in each case one pearl is superfluous. And if I divide it by 7, the pearls split evenly, but this is also impossible, as this necklace according to the covenant of ancestors must be passed to only one daughter-in-law. Therefore I will pass the necklace to whom who will determine how many pearls are in this necklace. Others should not be offended."

"It is known that the daughter-in-law, who decided this problem called Alkhan-Tumar, which means Necklace-Mascot.

Legend also says that this necklace still exists."

Naturally today it is not difficult to solve the problem. Let's first recall Wilson's theorem which is formulated as: a natural number n > 1 is a prime number if and only if $(n-1)!+1$ is divided by n evenly [1] .

This formulation implies that $(n-1)!+1$ is divided by all natural numbers less than n (except 1) with a remainder of 1. Using given theorem, let's find a solution: 721. However, it is not a full solution and it is one of a set of solutions.

Using criterions for divisibility and properties of natural numbers factorial expansion, we can find a first solution as 301. Obviously, that the solutions of this problem set up an arithmetic progression, the first term of which is equal to 301 and the progression difference is 420. i.e. the sequence of solutions of this problem looks like: 301, 721, 1141, etc.

This example is interesting because, if in this problem we replace last number 7 by number 9, or by any composite number, then this problem has no solution, since a condition of a remainder of 1 will not be fulfilled. In short, this problem has a solution when and only when a final number is a prime number, such as 11, 13, $17,\cdots,101,\cdots,211,\cdots$

As you can see, this problem is devoted to the problems of prime numbers. We believe that such problems with some similar formulations can be found in folklores of many nations. This is not surprising, the problems of prime numbers appeared before the Common Era, have been affecting interests of the scientific community for more than 2300 years. Since Eratosthenes, scientists have been gradually progressing, and in recent decades computers appeared to help them. But the main problems of prime numbers are still unsolved.

The solution of the above mentioned problem shows a way for solving the following problem of prime numbers.

2. PRIME NUMBERS GENERATION

Let we solve the following problem.

Suppose we are given an ordered sequence of prime numbers. It is necessary to find a next in order prime number. To solve the problem the following theorem is suggested.

Theorem. If the numbers $p_1, p_2, \cdots, p_i, \cdots, p_n$ are terms of the original sequence of prime numbers, where is an order of a prime number, i.e. $p_1 = 2$, $p_2 = 3$, $p_3 = 5$, $p_4 = 7$, etc., then there is a whole number k for which $p_1 \times p_2 \times \cdots \times p_{n-1} \times k + 1$ is divided evenly by p_n, and by p_i with a remainder of 1 for all $i = 1, 2, \cdots, n-1$.

It is easy to prove this theorem. For this purpose, it is only necessary to combine Wilson's theorem with a fundamental theorem of arithmetic ("every natural number greater than 1 can be represented as a product of prime numbers, and this product is unique"). The combination of these theorems makes required proof elementary and obvious. For this purpose, in a first approximation, it suffices to take as a product of all composite numbers less than p_n, i.e. a product of those composite numbers that appear in Wilson's theorem. Therefore, we skip a proof of the theorem intentionally.

Now, using this theorem, we will show an algorithm for solving posed problem.

3. ALGORITHM FOR GENERATING PRIME NUMBERS

Suppose we have a sequence of known prime numbers $p_1, p_2, \cdots, p_i, \cdots, p_n$, where i is an order number of prime numbers, i.e. $p_1 = 2$, $p_2 = 3$, $p_3 = 5$, $p_4 = 7$, etc. and p_n desired prime number.

Let indicate P_{n-1} as a product of $p_1 \times p_2 \times \cdots \times p_{n-1}$. Therefore, according to the conditions of the suggested theorem, for some whole number $k > 0$ the following equality should be satisfied:

$$\frac{P_{n-1} \times k + 1}{p_n} = \frac{(p_1 \times p_2 \times \cdots \times p_{n-1}) \times k + 1}{p_n} = \text{a whole number}, \quad (1)$$

Lemma. If equality (1) is true, then it is true for a infinite set of whole values of k. Plus a sequence of values of k forms an arithmetic progression, the first member of which is in interval from 1 to p_n, a common difference of this progression is equal to p_n, and one of the terms of the progression is a product of all composite numbers less p_n, i.e. a product of those composite numbers that appear in Wilson Theorem.

4. PROOF

For optimality of the proposed theorem implementation in practice, we need to determine a minimum, i.e. the initial value, of k or at least to identify an interval it belongs to.

For this we consider the ratio of $\frac{P_{n-1} \times k + 1}{p_n}$.

First we will show, that, if the ratio at any k from 1 to p_n is not a whole number, then it will not be a whole number for any $k = 1,2,3,\cdots,\infty$. Let prove this.

Suppose an expression $\frac{P_{n-1} \times k_1 + 1}{p_n}$ at $k_1 = 1,2,\cdots,p_n$ is not a whole number. Let try to determine, what this ratio will be for the next following values of k For this purpose we select a value of $k = k_1 + p_n$. Then the ratio in question takes a form as:

$$\frac{P_{n-1} \times (k_1 + p_n) + 1}{p_n} = \frac{P_{n-1} \times p_n}{p_n} + \frac{P_{n-1} \times k_1 + 1}{p_n}$$

It is obvious, that this expression will not be a whole number, since an addend, as mentioned above, is not a whole number. After that, repeating this procedure similarly for any $k = k_1 + i \times p_n$, where $i = 0,1,2,\cdots,\infty$, we assure that this ratio will never be a whole number.

This means that, if the ratio in question of Equation (1) is a whole number, then the initial value of k must be in interval from 1 to p_{n-1}, i.e. if

$$\frac{P_{n-1} \times k_1 + 1}{p_n} = \text{a whole number, then } k_1 > 0 \text{ and } k_1 < p_n$$

We will also show here that when the equality (1) is satisfied, the parameter k takes a set of values that equal to $k = k_1 + i \times p_n$, where $i = 0,1,2,\cdots,\infty$. In fact, using this expression, we transform the original ratio in question into:

$$\frac{P_{n-1} \times k + 1}{p_n} = \frac{P_{n-1} \times i \times p_n}{p_n} + \frac{P_{n-1} \times k_1 + 1}{p_n}$$

It is obvious, that the first term is a whole number and the second term, as was shown above, is also a whole number, therefore the ratio in question in the lump is also a whole number.

From the above said it follows that under the equality (1), the parameter k takes is a set of values which, as stated above, form an arithmetic progression. The first term of the progression should be in interval from 1 to p_{n-1}, i.e. $0 < k < p_n$. The common difference of this progression is equal p_n. One of the terms of the arithmetic progression is a product of all composite numbers less than p_n, i.e. it is equal to the product of those composite numbers that appear in Wilson's theorem.

5. THE PROOF OF THE LEMMA IS COMPLETED

From this moment and further, it is sufficient to know an initial value of parameter k, which is in interval of $0 < k < p_n$. For convenience, we introduce parameter $P(n,k)$, such that $P(n,k) = P_{n-1} \times k + 1 = (p_1 \times p_2 \times \cdots \times p_{n-1}) \times k + 1$.

Note that expression of Equation (1) includes two conditions:

a) A number $P_{n-1} \times k+1$ should be separately divided by all $p_1, p_2, \cdots, p_{n-1}$ with a remainder 1, i.e. for all $i=1,2,\cdots,n-1$:

$$P(n,k)=1(\bmod p_i) \tag{1a}$$

b) A number $P_{n-1} \times k+1$ should be divided by p_n evenly, i.e.

$$P(n,k)=0(\bmod p_n) \tag{1b}$$

Thus, let we have an initial ordered sequence of prime numbers $p_1, p_2, \cdots, p_i, \cdots, p_{n-1}$ To find the next in order prime number p_n we will build the following algorithm. For this, first for p_n we take the next odd number to p_{n-1}. After that, taking k from 1 to p_{n-1}, we execute arithmetic operation given in Equation (1). If in this case and at some value of k the conditions of Equations (1a) and (1b) will be observed, i.e. a result of division according the Equation (1) will be a whole number and then considered number p_n is the next prime number. After that, in the same way we start to search next prime number p_{n+1}.

And, if the conditions of Equations (1a) and (1b) are not met, then we deal with a composite number, i.e. in this case, considered number of p_n is not a prime number. It should be noted here that if the number p_n will be a composite number, then during the calculation of $\frac{P_{n-1} \times k+1}{p_n}$, a sequence of digits of the fractional part of this division result is periodically repeated while sorting parameter k. It can be easily seen after a few steps of cycle by k parameter, and it becomes clear that considered number p_n is not a prime number. Then we can immediately stop the cycle. This greatly increases the efficiency and speed of the proposed algorithm.

Then we proceed to the next odd number. The procedure is repeated again for different values of k.

If even in this case, the selected number is again a composite number, then proceed again to the next odd number. The operation is repeated until the conditions of Equations (1a) and (1b) will not be fulfilled. For clarity, we will show this based on a simple example.

Suppose we have an original sequence of prime numbers $p_1 = 2, p_2 = 3$, $p_3 = 5$ and $p_4 = 7$. It is required to find the next fifth prime number $p_5 = ?$ In this case, $n = 5$. For this, first for p_n we take 9 as an odd number next to 7. Then we calculate the values of $P(n,k) = P(5,k)$ for different values of k from 1 to $P(5,1) = 211,\ p_{n-1}$:

$$P(5,2) = 421,\ P(5,3) = 631,\ P(5,4) = 841,\ P(5,5) = 1051,\ P(5,6) = 1261$$

, etc. As can be seen, the values of $P(n,k)$ constitutes an arithmetic progression with a difference of $= 2 \times 3 \times 5 \times 7 = 210$. Note that among these values 841 and 1261 are composite numbers, and the rest of them are prime numbers. But we are looking for a prime number following 7.

Conducting the series of calculations, we see that a result of division of $\frac{P(5,k)}{9}$ at any value of $k<9$ will not be a whole number. This means that 9 is a composite number. After that for p_n we take the next odd number 11. Repeating the operation we obtain that at k = 10 value of $P(n,k)=P(5,10)=2101$ is divided by 11 evenly. Actually, 2101:11 = 191. This means that a prime number next to 7 is 11, i.e. $p_n = p_5 = 11$.

For completeness of the visualization, we consider one more sequence of prime numbers, with its last term as 23. In this case, first for p_n we take 25 as an odd number, next to number 23. And we see that it is a composite number. Then we select a number 27. At this time, the condition of Equation (1) will not be fulfilled, that is, we see that 27 is a composite number. When we select number 29, then at k = 17 condition of Equation (1) is satisfied, i.e. a value of $\frac{P(n,k)}{29}=\frac{P(10,17)}{29}=(223,092,870\times17+1)/29=130,778,579$ is a whole number. This means that a prime number after 23 will be 29, i.e. $p_n = p_{10} = 29$.

Sometimes it happens, that even at k = 1, we obtain desired results. For example, a value of $\frac{P(n,k)}{19}=\frac{P(8,1)}{19}=\frac{(51010\times1+1)}{19}=26869$ is a whole number, i.e. 19 is a prime number after 17. In addition, the numbers 17 and 19 are twins.

Some answers to the posed problem for a small set of primary prime numbers are given as Table 1. In this table, n is a counting number of a prime number.

In short, if there is a set of primary prime numbers, then while using expression of Equation (1) it is always possible to find the next prime number.

To show diversity to the proposed method, we offer one more elegant algorithm, the essence of which is primarily to find a value of k through existing prime numbers.

For this, we use Chinese remainder theorem, which states: "If natural numbers $p_1, p_2, \cdots, p_n$ are coprimes in pairs, then for any whole numbers $r_1, r_2, \cdots, r_n$ such that $0 \le r_i < p_i$ for all $i \in (1,2,3,\cdots,n)$ there is a number N, which when divided by p_i gives a reminder r_i for all $i \in (1,2,3,\cdots,n)$." [2].

It is known that constructive method for proving this theorem allows to solve the following system of linear equations modulo [3] [4]:

$$
\begin{aligned}
x &= r_1 \pmod{p_1} \\
x &= r_2 \pmod{p_2} \\
& \\
& \\
x &= r_{n-1} \pmod{p_{n-1}} \\
x &= r_n \pmod{p_n}
\end{aligned}
\qquad (2)
$$

If sets $(p_1, p_2, \cdots, p_n)$ и $(r_1, r_2, \cdots, r_n)$ fulfill conditions of Chinese theorem, then solution for system of Equations (2) exists and is unique within the accuracy of an operation by modulo $P_n = \prod_{i=1}^{n} p_i = p_1 \times p_2 \times \cdots \times p_n$, and this solution looks like [2] - [4]:

$$x = \sum_{i=1}^{n} r_i M_i M_i^{-1} \tag{3}$$

where

$$M_i = p_n / p_i \tag{4}$$

$$M_i M_i^{-1} = 1 \pmod{p_i} \tag{5}$$

i.e. M_i^{-1} is inverse for M_i by module p_i.

Table 1. Using with proposed theorem found primes number.

n	p_n	k	$P_{n-1} \times k + 1$
1	3	1	3
2	5	4	25
3	7	3	91
4	11	10	2101
5	13	10	23101
6	17	2	60061
7	19	1	510511

Now, knowing the solution of Equation (3) we can easily find the values of factor k. It is obvious, that Chinese theorem is true for any sequence of prime numbers since prime numbers are always coprimes in pairs. Then, considering for $p_1, p_2, \cdots, p_n$ in the system of Equation (2) a sequence of prime numbers, for our case we obtain that

$$r_1 = r_2 = \cdots = r_{n-1} = 1 \text{ and } r_n = 0 \qquad (6)$$

Now, by combining the proposed theorem with Chinese theorem, we have from Equations (1)-(6):

$$k = p_n \left(\sum_{i=1}^{n-1} \frac{M_i^{-1}}{p_i} - \frac{1}{P_n} \right) \qquad (7)$$

Expression of Equation (7) shows that if a sequence of prime numbers is known, we are always able to calculate a value of k in advance. Then, by substituting of calculated value of k in Equation (1), we verify the fulfillment of the conditions of Equations (1a) and (1b). If these conditions are not fulfilled, then a number considered as p_n is not a prime number. This number is the next composite number that stands for the prime number p_{n-1}. And, if the conditions of Equations (1a) and (1b) are met, then p_n is a desired prime number. After that, we proceed with finding next prime p_{n+1} and the cycle repeats.

In this case algorithm for searching prime numbers in odd numbers series looks as follows.

Step 1. Input values of existing prime numbers $p_1, p_2, \cdots, p_{n-1}$, where $p_1 = 2,\ p_2 = 3,\ p_3 = 5$, etc.

Step 2. For p_n, we input an odd number, which in a line of odd numbers follows a prime number p_{n-1}. Here $p_n = p_{n-1} + 2m$, where $m = 1, 2, 3, \cdots$ In this case for 1st step $m = 1$.

Step 3. Calculate $P_{n-1} = \prod_{i=1}^{n-1} p_i$ and $P_n = P_{n-1} \times p_n$.

Step 4. Calculate $M_i = p_n / p_i$ for all $i \in (1, 2, 3, \cdots, n)$.

Step 5. Using extended Euclid's algorithm, we find M_i^{-1} for all $i \in (1, 2, 3, \cdots, n)$ from the condition $M_i^{-1} M_i = 1 (\bmod p_i)$

Step 6. Calculate the desired value of k by the formula:

$$k = p_n \left(\sum_{i=1}^{n-1} \frac{M_i^{-1}}{p_i} - \frac{1}{P_n} \right) \bmod P_n$$

Step 7. Check fulfillment of equality (1).

Step 8. Conditional operator works here.

-If conditions of Equations (1a) and (1b) are not met, then considered number p_n is not a prime number, then assign m = m + 1 and proceed to step 2. The cycle repeats until conditions of Equations (1a) and (1b) will not be met.

-If conditions of Equations (1a) and (1b) are fulfilled, then p_n is a desired prime number, followed a prime number p_{n-1}.

Table 2. Comparison results between Wilson's theorem and proposed theorem.

Results obtained based on Wilson's theorem			Results obtained based on proposed theorem			
n	$(P_{n-1} \times k+1) \bmod p_n$	$((n-1)!+1) \bmod n$	p_i	$P_{n-1} = p_1 \cdot p_2, \cdots, p_{n-1}$	k	$(P_{n-1} \times k+1) \bmod p_n$
1	2	3	1	2	3	4
2	1	0	2	1	1	0
3	2	0	3	2	1	0
4	6	3				
5	24	0	5	6	4	0
6	120	1				
7	720	0	7	30	3	0
8	5,040	1				
9	40,320	1				
10	362,880	1				
11	3,628,800	0	**11**	210	10	0
12	39.916.800	1				
13	479,001,600	0	**13**	2310	10	0
14	6,227,020.800	1				
15	87,178,291,200	1				
16	1.307,674,368,000	1				
17	20,922,789,888,000	0	**17**	30,030	2	0
18	355,687,428,096,000	1				
19	6,402,373,705,728,000	0	**19**	510,510	1	0
20	121,645,100,408,832,000	1				
21	2.432,902,008,176,640,000	1				
22	51,090,942,171,709,400,000	1				
23	1,124,000,727,777,610,000,000	0	**23**	9,699,690	3	0
24	25,852,016,738,885,000,000,000	1				
25	620,448,401,733,239,000,000,000	1				
26	15,511,210,043,331,000,000,000,000	1				
27	403,291,461,126,606,000,000,000,000	1				
28	10,888,869,450,418,400,000,000,000,000	1				
29	304.888.344,611,714,000,000,000,000,000	0	**29**	223,092,870	17	0
30	8.841,761,993,739,700,000,000,000,000,000	1				
31	265,252,859,812,191,000,000,000,000,000,000	0	**31**	6,469,693,230	13	0
32	8,222,838,654,177,920,000,000,000,000,000,000	1				
33	263,130,836,933,694,000,000,000,000,000,000,000	1				
34	8,683,317,618,811,890,000,000,000,000,000,000,000	1				
35	295,232,799,039,604,000,000,000,000,000,000,000,000	1				
36	10,333,147,966,386,100,000,000,000,000,000,000,000,000	1				
37	371,993,326,789,901,000,000,000,000,000,000,000,000,000	0	**37**	200,560,490,130	10	0
38	13,763,753,091,226,300,000,000,000,000,000,000,000,000,000	1				
39	523,022,617,466,601,000,000,000,000,000,000,000,000,000,000	1				
40	20,397,882,081,197,400,000,000,000,000,000,000,000,000,000,000	1				
41	815,915,283,247,898,000,000,000,000,000,000,000,000,000,000,000	0	**41**	7,420,738,134,810	34	0

Step 9. Generation of the next prime number p_{n+1} is carried out similarly.

The cycle repeats.

We assume that the process of generating prime numbers based on the proposed theorem is faster than a generation based on Wilson's theorem.

In case of Wilson's theorem it is quite difficult to calculate the factorial of (n – 1)!. In fact, if generation is carried out in the area of the large numbers, then calculation of given factorial creates significant difficulties. This is explained by the fact that there are intervals in the sequence of natural numbers which include thousands, millions, billions and even some arbitrarily large number of natural numbers standing in a row, among which there is no prime number. For example, if you set an arbitrary large natural number m, let build a series of numbers $m!+2,\ m!+3,\ m!+4,\cdots,\ m!+m$ then it is obvious that each of these numbers is a composite number. You can easily check, particularly m! + 2 is evenly divided by 2, m! + 3 is divided by 3, and + is evenly divided by m, i.e. there is no a prime number in the large interval of $[(m!+2)/(m!+m)]$. If, for example, m = 10^{10}, then in case of Wilson's theorem, calculation of the factorial will inevitably lead to a large number of calculations involving a huge amount of large composite numbers. And in case of the proposed theorem calculations are mainly made with prime numbers.

This is clearly illustrated from Table 2, which shows results of searching prime numbers using methods following from Wilson's theorem (left part of the table) and proposed theorem (right part of the table).

The left part of the table provides calculations made for values of from 2 to 41. In the 1st column of left part of the table (case of Wilson's theorem) all natural numbers are given. In this column cells containing prime numbers which are green highlighted for illustrative purposes. In the 2nd column the calculated values of $(n-1)!$ factorial are shown. And 3rd column provides remainders which take zero values in case of prime numbers.

Similar results, obtained using proposed theorem, are shown in the right part of the table. First column of the right part shows prime numbers. Second column of this part presents calculated values of $P_{n-1} = \prod_{i=1}^{n-1} p_i$. And the last column shows values of remainders which also take zero values for prime numbers.

Table 2 shows that, in case of Wilson's theorem, in order to carry out generation of prime numbers at least up to 13th prime number, i.e. up to $p_{13} = 41$ it is necessary to perform a lot of laborious calculations with large numbers.

A right part of the table shows only those numbers with which the calculations have been made using the proposed method. Comparing the left and right parts of the table, we can see that efficiency, speed, and convenience of the proposed method are beyond question, this is obvious. Plus, for a set of large integers, this obviousness becomes even more than self-evident.

Note again that in case of Wilson's theorem the complexity lies in calculation of $(n-1)!$ factorial. It is easier to calculate a^{n-1}; therefore elementary tests, determining whether a number is prime number, are based on Fermat's theorem, rather than on Wilson's theorem. However, note that in contrast to Fermat's small theorem, Wilson's theorem simultaneously is a necessary and sufficient condition for determining primality of any number.

REFERENCES

1. Vinogradov, I.M. (1952) Fundamental of the Theory of Number. 5th Edition, Publishing House of Technology & Scientific Literature, 262.
2. Ishmuchametov, Sh.T. (2011) Methods of Factoring Natural Numbers. Kazan Federal University Press, Kazan, 202.
3. Nesterenko, A. (2011) Introduction to Modern Cryptography, Theoretical Numbers Algorithms. 190. http://img0. liveinternet. ru/images/attach/c/4/3908/3908902_ntheory.pdf
4. Gabidulin, E.M., Kshevetshkii, A.S. and Kolybelnikov, A.I. (2011) Information Security. MFTI, 262.

CHAPTER 2

On the k-Lucas Numbers of Arithmetic Indexes

Sergio Falcon

Department of Mathematics and Institute for Applied Microelectronics (IUMA), University of Las Palmas de Gran Canaria, Las Palmas, Spain

ABSTRACT

In this paper, we study the k-Lucas numbers of arithmetic indexes of the form an + r, where n is a natural number and r is less than r. We prove a formula for the sum of these numbers and particularly the sums of the first k-Lucas numbers, and then for the even and the odd k-Lucas numbers. Later, we find the generating function of these numbers. Below we prove these same formulas for the alternated k-

Lucas numbers. Then, we prove a relation between the k-Fibonacci numbers of indexes of the form $2^r n$ and the k-Lucas numbers of indexes multiple of 4. Finally, we find a formula for the sum of the square of the k-Fibonacci even numbers by mean of the k-Lucas numbers.

KEYWORDS

k-Fibonacci Numbers; k-Lucas Numbers; Generating Function

1. INTRODUCTION

Let us remember the k-Lucas numbers $L_{k,n}$ are defined [1] by the recurrence relation $L_{k,n+1} = k\,L_{k,n} + L_{k,n-1}$ with the initial conditions $L_{k,0} = 2,\ L_{k,1} = k$

Among other properties, the Binnet Identity establishes $L_{k,} = \sigma_1^n + \sigma_2^n$ being $\sigma_1 = \frac{k+\sqrt{k^2+4}}{2}$ and $\sigma_2 = \frac{k-\sqrt{k^2+4}}{2}$ the characteristic roots of the recurrence equation $r^2 - k\,r - 1 = 0$.

Evidently, $\sigma_1\sigma_2 = -1,\ \sigma_1 + \sigma_2 = k,\ \sigma_1 - \sigma_2 = \sqrt{k^2+4}$.

Moreover, it is verified [1, Theorem 2.4] that $L_{k,n} = F_{k,n-1} + F_{k,n+1}$.

If we apply iteratively the equation $L_{k,n+1} = k\,L_{k,n} + L_{k,n-1}$ then we will find a formula that relates the k–Lucas numbers to the k–Fibonacci numbers:

$$L_{k,n} = L_{k,n-(p-1)}F_{k,p} + L_{k,n-p}F_{k,p-1} \qquad (1.1)$$

This formula is similar to the Convolution formula for the k-Fibonacci numbers $F_{k,n+m} = F_{k,n+1}F_{k,m} + F_{k,n}F_{k,m-1}$ [2,3].

Moreover, we define $L_{k,-n} = (-1)^{n+1} L_{k,n}$. Then, if we do p = −n in Formula (1.1) obtain

$$L_{k,n} = (-1)^{n+1}\left(L_{k,2n+1}F_{k,n} - L_{k,2n}F_{k,n+1}\right).$$

2. ON THE K-LUCAS NUMBERS OF ARITHMETIC INDEX

We begin this section with a formula that relates each other some k-Lucas numbers.

2.1. Theorem 1 (The k-Lucas Numbers of Arithmetic Index)

If a is a nonnull natural number and r = 0, 1, 2, ... a − 1, then

$$L_{k,a(n+1)+r} = L_{k,a}L_{k,an+r} - (-1)^{a} L_{k,a(n-1)+r} \qquad (2.1)$$

Proof. In [4] it is proved

$$F_{k,a(n+1)+r} = \left(F_{k,a-1} + F_{k,a+1}\right) F_{k,an+r} - (-1)^a F_{k,a(n-1)+r}.$$

Then

$$L_{k,a(n+1)+r} = F_{k,a(n+1)+r-1} + F_{k,a(n+1)+r+1}$$
$$= \left(F_{k,a-1} + F_{k,a+1}\right) F_{k,an+r-1}$$
$$-(-1)^a F_{k,a(n-1)+r-1} + \left(F_{k,a-1} + F_{k,a+1}\right) F_{k,an+r+1}$$
$$-(-1)^a F_{k,a(n-1)+r+1}$$
$$= L_{k,a} L_{k,an+r} - (-1)^a L_{k,a(n-1)+r}$$

If r = 0, then $L_{k,a(n+1)} = L_{k,a} L_{k,an} - (-1)^a L_{k,a(n-1)}$

In this case, if a = 2p + 1, then an odd k-Lucas number can be expressed in the form

$$L_{k,(2p+1)(n+1)} = L_{k,2p+1} L_{k,(2p+1)n} + L_{k,(2p+1)(n-1)}$$

Applying iteratively Formula (2.1), the general term, for $0 \le m \le n$, can be written like a non-linear combination of the form

$$L_{k,an+r} = \sum_{j=0}^{m} (-1)^{j(a+1)} \binom{m}{j} L_{k,a}^{m-j} L_{k,a(n-m-j)+r}$$

In particular, if m = n, then

$$L_{k,an+r} = \sum_{j=0}^{n} (-1)^{j(a+1)} \binom{n}{j} L_{k,a}^{n-j} L_{k,r-aj}$$

2.2. Generating Function of the Sequence $\{L_{k,an+r}\}$

Let $l(k,x;a,r)$ be the generating function of the sequence $\{L_{k,an+r}\}$. That is,

$$l(k,x;a,r) = L_{k,r} + L_{k,a+r}x$$
$$+L_{k,a+2r}x^2 + L_{k,a+3r}x^3 + \cdots$$

Then,

$$L_{k,a}x\cdot l(k,x;a,r)=L_{k,a}L_{k,r}x+L_{k,a}L_{k,a+r}x^2$$
$$+L_{k,a}L_{k,a+2r}x^3+\cdots$$

And

$$(-\ )^a x^2 l(k\ x\ a\ r)$$
$$=(-1)^a L_{k,r}x^2+(-1)^a L_{k,a+r}x^3+\cdots$$

from where $l(k,x;a,r)\times\left(1-L_{k,a}x+(-1)^a x^2\right)=L_{k,r}+\left(L_{k,a+r}-L_{k,a}\right)x$ no more to take into account Formula (2.1). So, the generating function of the sequence $\left\{L_{k,an+r}\right\}$ is

$$l(k,x;a,r)=\frac{\left(L_{k,r}+L_{k,a+r}-L_{k,a}L_{k,r}\right)x}{1-L_{k,a}x+(-1)^a x^2}.$$

As particular case, if a = 1, then r = 0 and the generating function of the k-Lucas sequence $\{L_{k,n}\}$ is $l(k,x)=\frac{2-kx}{1-kx-x^2}$, that, for the classical Lucas sequence is $l(x)=\frac{2-x}{1-x-x^2}$

If we want to take out the two bisection sequences of the classical Lucas sequence (k = 1), the respective generating functions are a = 2 and r = 0: $l(x;2,0)=\frac{2-3x}{1-3x+x^2}$ that generates the sequence $\{L_{2n}\}=\{2,3,7,18,47,123,\cdots\}$ a = 2 and r = 1: $l(x;2,1)=\frac{1+x}{1-3x+x^2}$ that generates the sequence $\{L_{2n+1}\}=\{1,4,11,29,76,199,\cdots\}$.

2.3. Theorem 2 (Sum of the k-Lucas Numbers of Arithmetic Index)

If a is a nonnull natural number and r = 0, 1, 2, ... a − 1, then

$$\sum_{j=0}^{n} L_{k,aj+r}=\frac{L_{k,a(n+1)+r}-(-1)^a L_{k,an+r}+(-1)^r L_{k,a-r}-L_{k,r}}{L_{k,a}-(-1)^a-1} \qquad (2.2)$$

Proof.

$$\sum_{j=0}^{n} L_{k,aj+r} = \sum_{j=0}^{n}\left(\sigma_1^{aj+r} + \sigma_2^{aj+r}\right) = \frac{\sigma_1^{an+r} - \sigma_1^{r}}{\sigma_1^{a} - 1} + \frac{\sigma_2^{an+r} - \sigma_2^{r}}{\sigma_2^{a} - 1}$$

$$= \frac{(-1)^a \sigma_1^{an+r} - \sigma_1^{a(n+1)+r} - \sigma_1^{r}\sigma_2^{a} + \sigma_1^{r} + (-1)^a \sigma_2^{an+r} - \sigma_2^{a(n+1)+r} - \sigma_2^{r}\sigma_1^{a} + \sigma_1^{r}}{(-1)^a - \sigma_1^{a} - \sigma_2^{a} + 1}$$

$$= \frac{(-1)^a L_{k,an+r} - L_{k,a(n+1)+r} - (-1)^r L_{k,a-r} + L_{k,r}}{-L_{k,a} + (-1)^a + 1}$$

Because

$$\sigma_1^{r}\sigma_2^{a} + \sigma_1^{a}\sigma_2^{r} = \sigma_1^{r}\sigma_2^{a-r}\sigma_2^{r} + \sigma_1^{a-r}\sigma_1^{r}\sigma_2^{r} = (-1)^r\left(\sigma_2^{a-r} + \sigma_1^{a-r}\right)$$

and after applying the formula for the sum of a geometric progression.

2.4. Corollary 1 (Sum of Consecutive Odd k-Lucas Numbers)

If r = 0 and a = 2p + 1, Equation (2.2) is

$$\sum_{j=0}^{n} L_{k,(2p+1)j} = \frac{L_{k,(2p+1)(n+1)} + L_{k,(2p+1)n} - 2}{L_{k,2p+1}} + 1$$

In this case, the sum of the first k-Lucas numbers is (for p = 0),

$$\sum_{j=0}^{n} L_{k,j} = \frac{1}{k}\left(L_{k,n+1} + L_{k,n} - 2\right) + 1 \qquad (2.3)$$

that for the classical Lucas numbers is $\sum_{j=0}^{n} L_j = L_{n+2} - 1$

2.5. Corollary 2 (Sum of Consecutive Even k-Lucas Numbers)

If r = 0 and a = 2p, then Equation (2.2) is

$$\sum_{j=0}^{n} L_{k,2pj} = \frac{L_{k,2p(n+1)} - L_{k,2pn}}{L_{2p} - 2} + 1 \qquad (2.4)$$

In this case, if p = 1 we obtain the formula for the sum of the first even k-Lucas numbers $\sum_{j=0}^{n} L_{k,2j} = \frac{L_{k,2n+1}}{k} + 1$, and for the classical Lucas numbers is $\sum_{j=0}^{n} L_{2j} = L_{2n+1} + 1$

2.6. Theorem 3 (Sum of Alternated k-Lucas Numbers of Arithmetic Index)

For a > 0 and r = 0, 1, 2, ...a – 1, the sum of alternated k-Lucas numbers is

$$\sum_{j=0}^{n}(-1)^{j} L_{k,aj+r}$$

$$=\frac{(-1)^{a+n} L_{k,an+r}+(-1)^{n} L_{k,a(n+1)+r}+(-1)^{n} L_{k,a-r}+L_{k,r}}{L_{k,a}+(-1)^{a}+1}$$

Proof. As in the previous theorem,

$$\sum_{j=0}^{n}(-1)^{j} L_{k,aj+r}=\sum_{j=0}^{n}(-1)^{j}\left(\sigma_1^{aj+r}+\sigma_2^{aj+r}\right)$$

$$=\frac{(-1)^{n}\sigma_1^{an+r}\left(-\sigma_1^{a}\right)-\sigma_1^{r}}{-\sigma_1^{a}-1}+\frac{(-1)^{n}\sigma_2^{an+r}\left(-\sigma_2^{a}\right)-\sigma_2^{r}}{-\sigma_2^{a}-1}$$

$$=\frac{(-1)^{n}(-1)^{a}\sigma_1^{an+r}+(-1)^{n}\sigma_1^{a(n+1)+r}+(-1)^{r}\sigma_2^{a-r}+\sigma_1^{r}+(-1)^{n}(-1)^{a}\sigma_2^{an+r}+(-1)^{n}\sigma_2^{a(n+1)+r}+(-1)\sigma_1^{a-r}+\sigma_2^{r}}{(-1)^{a}+\sigma_1^{a}+\sigma_2^{a}+1}$$

$$=\frac{(-1)^{a}\left(\sigma_1^{an+r}+\sigma_2^{an+r}\right)+(-1)^{n}\left(\sigma_1^{a(n+1)+r}+\sigma_2^{a(n+1)+r}\right)+(-1)^{r}\left(\sigma_2^{a-r}+\sigma_1^{a-r}\right)+\sigma_1^{r}+\sigma_2^{r}}{L_{k,a}+(-1)^{a}+1}$$

$$=\frac{(-1)^{a+n} L_{k,an+r}+(-1)^{n} L_{k,a(n+1)+r}+(-1)^{n} L_{k,a-r}+L_{k,r}}{L_{k,a}+(-1)^{a}+1}$$

2.7. Corollary 3 (Sum of Consecutive Alternated Odd k-Lucas Numbers)

As particular case, if a = 2p + 1 and r = 0,

$$\sum_{j=0}^{n}(-1)^j L_{k,(2p+1)j}$$

$$=(-1)^n \frac{L_{k,(2p+1)(n+1)} - L_{k,(2p+1)n} + 2(-1)^n}{L_{k,2p+1}} + 1$$

Then, for p = 0 we obtain the sum of the first alternated k-Lucas numbers

$$\sum_{j=0}^{n}(-1)^j L_{k,2j+1} = (-1)^n \frac{L_{k,n+1} - L_{k,n} + 2(-1)^n}{k} + 1$$

, that for the classical Lucas numbers is

$$\sum_{j=0}^{n}(-1)^j L_{2j+1} = (-1)^n L_{k,n-1} + 3$$

.

2.8. Corollary 4 (Sum of Consecutive Alternated Even k-Lucas Numbers)

If r = 0 and a = 2p + 1, then

$$\sum_{j=0}^{n}(-1)^j L_{k,2pj}$$
$$=(-1)^n \frac{L_{k,2p(n+1)}+L_{k,2pn}+L_{k,2p}+2(-1)^n}{L_{k,2p}+2}$$

And for the first consecutive alternated even k-Lucas numbers

$$\sum_{j=0}^{n}(-1)^j L_{k,2j}=(-1)^n \frac{L_{k,2(n+1)}+L_{k,2n}+2(-1)^n+k^2+2}{k^2+4}$$ that for the classical Lucas numbers is

$$\sum_{j=0}^{n}(-1)^j L_{2j}=(-1)^n \frac{1}{5}\left(5F_{2n+1}+3+2(-1)^n\right).$$

3. ON THE K-FIBONACCI NUMBERS OF INDEXES N AND THE K-LUCAS NUMBERS

In this section we will study a relation between the numbers $F_{k,2^r n}$ and $L_{k,n}$.

3.1. Theorem 4 (A Relation between Some k-Fibonacci and the k-Lucas Numbers)

For r ≥ 1, it is

$$\frac{F_{k,2^r n}}{F_{k,n}} = \prod_{j=0}^{r-1} L_{k,2^j n} \tag{3.1}$$

Proof.

$$\frac{F_{k,2^r n}}{F_{k,n}} = \frac{\sigma_1^{2^r n} - \sigma_2^{2^r n}}{\sigma_1^n - \sigma_2^n} = \left(\sigma_1^{2^{r-1} n} + \sigma_2^{2^{r-1} n}\right)\frac{\sigma_1^{2^{r-1} n} - \sigma_2^{2^{r-1} n}}{\sigma_1^n - \sigma_2^n}$$

$$= L_{k,2^{r-1} n}\left(\sigma_1^{2^{r-2} n} + \sigma_2^{2^{r-2} n}\right)\frac{\sigma_1^{2^{r-2} n} - \sigma_2^{2^{r-2} n}}{\sigma_1^n - \sigma_2^n}$$

$$= L_{k,2^{r-1} n} L_{k,2^{r-2} n} \cdots \frac{\sigma_1^{2n} - \sigma_2^{2n}}{\sigma_1^n - \sigma_2^n} = \prod_{j=0}^{r-1} L_{k,2^j n}$$

In particular, if r = 1, it is $\frac{F_{k,2n}}{F_{k,n}} = L_{k,n}$

Taking into account $L_{k,m} = \sigma_1^m + \sigma_2^m$, if we expand Formula (3.1), we find that this formula can be expressed as $\frac{F_{k,2^r n}}{F_{k,n}} = \prod_{j=0}^{2^{r-1}-1} (-1)^{(j+1)n} L_{k,2(2j+1)n}$ or, that is the same,

$$\frac{F_{k,2^r n}}{F_{k,n}} = \prod_{j=0}^{2^{r-2}-1} L_{k,2(4j+3)n} + (-1)^n \sum_{j=0}^{2^{r-1}-1} L_{k,(4j+1)n}$$

Then, applying Formula (2.2) to the second hand right of this equation with $n = 2^{r-2} - 1$, a = 4n, and r = 3n for the first term and r = n for the second,

$$\begin{aligned}\frac{F_{k,2^r n}}{F_{k,n}} &= \frac{L_{k,(2^r+3)n} - L_{k,(2^r-1)n} + (-1)^n L_{k,n} - L_{k,3n}}{L_{k,4n} - 2} \\ &+ (-1)^n \frac{L_{k,(2^r+1)n} - L_{k,(2^r-3)n} + (-1)^n L_{k,3n} - L_{k,n}}{L_{k,4n} - 2} \\ &= \frac{L_{k,(2^r+3)n} - L_{k,(2^r-1)n} + (-1)^n \left(L_{k,(2^r+1)n} - L_{k,(2^r-3)n} \right)}{L_{k,4n} - 2}\end{aligned} \quad (3.2)$$

We tray to simplify the second hand right of this equation. For that, we will prove the following Lemma.

3.2. Lemma 1

$$L_{k,(a+4)n} - L_{k,an} = \left(k^2 + 4\right) F_{k,(a+2)n} F_{k,2n} \qquad (3)$$

Proof. We will apply the following formulas:

$L_{k,p} = F_{k,p-1} + F_{k,p+1}$ (relation)

$F_{k,-p} = (-1)^{p+1} F_{k,p}$ (negative)

$F_{k,a+b} = F_{k,a} F_{k,b-1} + F_{k,a+1} F_{k,b}$ (convolution)

$F_{k,p+1} = k\, F_{k,p} + F_{k,p-1}$ (definition)

Then:

$$L_{k,(a+4)n} - L_{k,an}$$
$$= F_{k,(a+4)n-1} + F_{k,(a+4)n+1} - F_{k,an-1} - F_{k,an+1} \text{ (by relation)}$$

$$
\begin{aligned}
&= F_{k,(a+2)n+(2n-1)} + F_{k,(a+2)n+(2n+1)} \\
&- F_{k,(a+2)n+(-2n-1)} - F_{k,(a+2)n+(-2n+1)} \\
&= F_{k,(a+2)n} F_{k,2n-2} + F_{k,(a+2)n+1} F_{k,2n-1} \\
&+ F_{k,(a+2)n} F_{k,2n} + F_{k,(a+2)n+1} F_{k,2n+1} \\
&- F_{k,(a+2)n} F_{k,-2n-2} - F_{k,(a+2)n+1} F_{k,-2n-1} \\
&- F_{k,(a+2)n} F_{k,-2n} - F_{k,(a+2)n+1} F_{k,-2n+1} \qquad \text{(by convolution)}
\end{aligned}
$$

$$
\begin{aligned}
&= F_{k,(a+2)n} F_{k,2n-2} + F_{k,(a+2)n+1} F_{k,2n-1} \\
&+ F_{k,(a+2)n} F_{k,2n} + F_{k,(a+2)n+1} F_{k,2n+1} \\
&+ F_{k,(a+2)n} F_{k,2n+2} - F_{k,(a+2)n+1} F_{k,2n+1} \\
&+ F_{k,(a+2)n} F_{k,2n} - F_{k,(a+2)n+1} F_{k,2n-1} \qquad \text{(by negative)}
\end{aligned}
$$

$$
\begin{aligned}
&= F_{k,(a+2)n} \left(F_{k,2n} - k\, F_{k,2n-1} + F_{k,2n} + k\, F_{k,2n+1} + F_{k,2n} \right) \\
&= F_{k,(a+2)n} \left(k^2 + 4 \right) F_{k,2n}
\end{aligned}
$$

(by definition)

And applying this Lemma to Equation (3.2), we will have:

$$\frac{F_{k,2^r n}}{F_{k,n}}$$
$$=\frac{\left(k^2+4\right)F_{k,\left(2^r+1\right)n}F_{k,2n}+(-1)^n\left(k^2+4\right)F_{k,\left(2^r-1\right)n}F_{k,2n}}{L_{k,4n}-2}$$

that is

$$\frac{F_{k,2^r n}}{F_{k,n}}=\frac{\left(k^2+4\right)F_{k,2n}}{L_{k,4n}-2}\left(F_{k,\left(2^r+1\right)n}+(-1)^n F_{k,\left(2^r-1\right)n}\right)$$

from where

$$\frac{F_{k,\left(2^r+1\right)n}+(-1)^n F_{k,(2^r-1)n}}{F_{k,2^r n}}=\frac{L_{k,4n}-2}{\left(k^2+4\right)F_{k,n}F_{k,2n}}$$
$$=\frac{L_{k,4n}-2}{\left(\sigma_1^n-\sigma_2^n\right)\left(\sigma_1^{2n}-\sigma_2^{2n}\right)}$$
$$=\frac{L_{k,4n}-2}{\left(\sigma_1^{3n}+\sigma_2^{3n}\right)-(-1)^n\left(\sigma_1^n+\sigma_2^n\right)}=\frac{L_{k,4n}-2}{L_{k,3n}-(-1)^n L_{k,n}}$$

If in Equation (3.3) it is a = 0, then it is $F_{k,2n}^2=\frac{L_{k,4n}-2}{k^2+4}$, and applying the Formulas (2.5) and (2.4),

$$\sum_{j=0}^{n} F_{k,2j}^2 = \frac{1}{k^2+4}\sum_{j=0}^{n}\left(L_{k,4j}-2\right)$$
$$= \frac{1}{k^2+4}\left(\frac{L_{k,4(n+1)}-L_{k,4n}}{L_{k,4}-2}+1-2(n+1)\right)$$

That is

$$\sum_{j=0}^{n} F_{k,2j}^2 = \frac{1}{k^2+4}\left(\frac{k\,L_{k,4n+3}+L_{k,4n+1}}{k\left(k^2+4\right)}-2n-1\right)$$

In particular, for the classical Lucas numbers (k = 1), it is $\sum_{j=0}^{n} F_{2j}^2 = \frac{1}{5}\left(\frac{L_{4n+3}+L_{4n+1}}{5}-2n-1\right)$.

4. ACKNOWLEDGEMENTS

This work has been supported in part by CICYT Project number MTM200805866-C03-02 from Ministerio de Educación y Ciencia of Spain.

REFERENCES

1. S. Falcon, "On the k-Lucas Numbers," International Journal of Contemporary Mathematical Sciences, Vol. 6, No. 21, 2011, pp. 1039-1050

2. S. Falcon and A. Plaza, "On the Fibonacci k-Numbers," Chaos, Solitons & Fractals, Vol. 32, No. 5, 2007, pp. 1615-1624.

3. S. Falcon and A. Plaza, "The k-Fibonacci Sequence and the Pascal 2-Triangle," Chaos, Solitons & Fractals, Vol. 33, No. 1, 2007, pp. 38-49.

4. S. Falcon and A. Plaza, "On k-Fibonacci Numbers of Arithmetic Indexes," Applied Mathematics and Computation, Vol. 208, 2009, pp. 180-185

CHAPTER 3

Composition of complex numbers: Delineating the computational role of the left anterior temporal lobe

Esti Blanco-Elorrieta a,*,c, Liina Pylkkänen a,b,c

a NYUAD Institute, New York University Abu Dhabi, Abu Dhabi, United Arab Emirates

b Department of Linguistics, New York University, New York, NY, USA

c Department of Psychology, New York University, New York, NY, USA

ABSTRACT

What is the neurobiological basis of our ability to create complex messages with language? Results from multiple methodologies have converged on a set of brain regions as relevant for this general process, but the computational details of these areas remain to be characterized. The left anterior temporal lobe (LATL) has been a consistent node within this network, with results suggesting that although it rather systematically shows increased activation for semantically complex structured stimuli, this effect does not extend to number phrases such as 'three books.' In the present work we used magnetoencephalography to investigate whether numbers in general are an invalid input to the combinatory operations housed in the LATL or whether the lack of LATL engagement for stimuli such as 'three books' is due to the quantificational nature of such phrases. As a relevant test case, we employed complex number terms such as 'twenty-three', where one number term is not a quantifier of the other but rather, the two terms form a type of complex concept. In a number naming paradigm, participants viewed rows of numbers and depending on task instruction, named them as complex number terms ('twenty-three'), numerical quantifications ('two threes'), adjectival modifications ('blue threes') or non-combinatory lists (e.g., 'two, three'). While quantificational phrases failed to engage the LATL as compared to non-combinatory controls, both complex number terms and adjectival modifications elicited a reliable activity increase in the LATL. Our results show that while the LATL does not participate in the enumeration of tokens within a set, exemplified by the quantificational phrases, it does support conceptual combination, including the composition of complex number concepts.

KEYWORDS

Conceptual combination; Magnetoencephalography; Left anterior temporal lobe; Numerical quantification; Complex number composition

1. INTRODUCTION

Understanding the brain basis of linguistic creativity is a fundamental goal for the cognitive neuroscience of language: what is the neurobiology of our ability to create an infinity of conceptual representations from the basic building blocks of language? Large networks of brain areas have been proposed to partake in the brain's "semantic network" (Binder et al., 2009 and Binder and Desai, 2011) including the left inferior frontal cortex (e.g., Hagoort and Indefrey, 2014), the superior temporal gyrus (e.g., Friederici, 2011), the angular gyrus (e.g., Price et al., 2015) and the left anterior temporal lobe (LATL). Each of these regions has been proposed to carry a role in the combinatory processing of language. Damage to the angular gyrus can result in a wide variety of neuropsychological conditions affecting language, visuo-spatial processing and number cognition and thus it has been proposed as a high-level supramodal integration area, with the combination of concepts as part of its computational profile (Binder et al., 2009). The anatomical connectivity of the angular gyrus further conforms to a high level integrative role as it receives its input mostly from other association areas as opposed to primary sensory cortices (Bonner et al., 2013, Mesulam, 2000, Pandya and Seltzer, 1982 and Yeterian and Pandya, 1985). The left inferior frontal cortex has also been associated with a multitude of functions, including phonological (Heim et al., 2008), semantic (Thompson-Schill et al., 1997) and syntactic processing (Stromswold et al., 1996), but within combinatory processing, its contribution has most commonly been proposed to be syntactic (Indefrey, 2012, Indefrey et al., 2001b, Hagoort and Indefrey, 2014, Friederici, 2011, Pallier et al., 2011 and Tyler et al., 2011). Similar sensitivity to syntactic stimulus properties has been observed in posterior superior temporal cortex (Hagoort and Indefrey, 2014 and Pallier et al., 2011).

However, as regards the semantic aspects of combinatory processing, multiple methodologies, including neuroimaging, electrophysiology and patient research, have produced an internally highly consistent body of work strongly implicating the LATL as a basic site for semantic combination. Core evidence for this include hemodynamic and neuropsychological research proposing that this brain area acts as a 'semantic hub' in which conceptual representations are bound together and processed by a common set of neurons (Bright et al., 2004, Clarke et al., 2011, Clarke et al., 2013, Gauthier et al., 1997,Grabowski et al., 2001, Rogers et al., 2006 and Tyler et al., 2004) as well as sentence processing studies showing that structured sentences elicit greater LATL activity than meaningless sentences or word lists (Friederici et al., 2000, Humphries et al., 2006,Humphries et al., 2007, Mazoyer et al., 1993, Pallier et al., 2011, Rogalsky and Hickok, 2009, Stowe et al., 1998, Vandenberghe et al., 2002 and Xu et al., 2005). More recently, magnetoencephalography (MEG) studies on minimal combinations of two words have demonstrated that this activity relates to very basic combinatory operations as opposed to sentence-level phenomena both in comprehension (Bemis and Pylkkänen, 2011 and Bemis and Pylkkänen, 2012) and in production (Del Prato and Pylkkänen, 2014 and Pylkkänen et al., 2014).

While this large dataset on the LATL is still compatible with many definitions of "semantic processing", the robustness of these findings and their generality across multiple methodologies presents an opportunity for a systematic investigation of the computational details of this activity. One step towards sharpening our understanding involves recent MEG results on language production (Del Prato and Pylkkänen, 2014), where the modification of object denoting nouns with color adjectives (*blue cups*) engaged the LATL, while numerical quantification of the same nouns (*two cups*) did not. Given that both of these combinations involve semantic composition, these data are incompatible with a general semantic composition account of the LATL. Instead, they suggest a narrower computation, perhaps better characterized as a type of "conceptual combination", a label employed in the concepts and categories literature for a host of cases where, intuitively, the combination of two concepts serves to form a more complex one, typical examples being adjective–noun and noun–noun combinations. Given that in phrases such as *two cups*, *two* does not add a feature to the concept denoted by *cup* but rather enumerates the number of tokens in a set of cups, such cases would, by hypothesis, fall outside the definition of conceptual combination that is relevant for the LATL. Related evidence for the conceptual nature of the LATL include the sensitivity of its combinatory response to conceptual specificity (Westerlund and Pylkkänen, 2014 and Zhang and Pylkkänen, 2015) and the correlation between the LATL activation elicited by specific concepts like *boy* and the product of the activations for their constituent concepts (i.e., *male* and *child*) (Baron and Osherson, 2011).

The purpose of the current experiment was to further characterize which input elements and specific computations constitute the "conceptual combinations" which drive activity within the LATL. Specifically, our study was designed around the question of whether complex number terms, such as *thirty-two*, would elicit combinatory activity in the LATL, despite its insensitivity to numerical quantification. Since this study builds on the results ofDel Prato and Pylkkänen (2014), which was conducted in production, the current study is also a production study. Several prior studies have addressed the neurobiological similarity of combinatory operations in production vs. comprehension, with results compellingly showing that similar regions are recruited for composition whether the participant is comprehending or producing language (Hagoort and Indefrey, 2014,Menenti et al., 2011, Segaert et al., 2012 and Pylkkänen et al., 2014). On the basis of this, one would predict the results of the current study to be replicable in comprehension. Further, as described in Methods, our production paradigm allowed us to keep the physical stimulus almost completely constant across conditions (cf., Del Prato and Pylkkänen,

2014 and Pylkkänen et al., 2014), which was particularly useful given that confounding low level factors are often an issue in language studies. The combination of our two-word paradigm together with the millisecond time-resolution of MEG circumvents the principle obstacle behind electrophysiological investigations of sentence production, i.e., that meaningful electrophysiological data is extremely difficult to collect while the mouth is moving. However, the syntactic and semantic planning of small two-word phrases is thought to occur entirely prior to the onset of articulation (Alario et al., 2002,Meyer, 1996 and Schriefers et al., 1999) and thus with a technique capable of capturing these planning stages millisecond-by-millisecond, we are able to measure combinatory processing (Pylkkänen et al., 2014). An added advantage of the detailed time resolution is that it allows us to separate different effects within the same region at different times.

Behavioral research on conceptual combination has classically been quite focused on one particular domain; the modification of nouns (e.g., Medin and Shoben, 1988, Murphy, 1990, Wisniewski, 1996 and Hampton, 1997). Given that the LATL is at least activated by the core cases of conceptual combination, as evidenced by the many studies on adjective–noun combinations (comprehension: Bemis and Pylkkänen, 2013 and Westerlund and Pylkkänen, 2014 and production: Del Prato and Pylkkänen, 2014 and Pylkkänen et al., 2014), it now becomes possible to concretely test what types of semantic combinations drive this activity. In other words, what is the brain's definition of "conceptual combination"?

Number words are a particularly interesting test case for this purpose as they are a very multifaceted word class in terms of the position and semantic functions they can fulfill in a sentence (Hurford, 1975). The most widely spread view states that (simplex) cardinals such as 'one', 'two' and 'three' are determiners (Barwise and Cooper, 1981, Bennett, 1975, Montague, 1974 and Scha, 1984) and they have traditionally been treated either as generalized quantifiers (Montague, 1974 and Barwise and Cooper, 1981) or restrictive modifiers (Link, 2002) when they precede the noun. However, according to Hurford, 1975, Hurford, 1987, Hurford, 2001 and Hurford, 2003 and Ionin and Matushansky (2006): "when not acting as modifiers, the vast majority of simplex cardinals are singular nouns and belong to one or another open lexical class available in a language". Therefore, number words do not fall clearly in either open or close class word categories and can interestingly occupy the place of both in a noun phrase. This unique feature provides the opportunity to create different combinations and investigate to which extent the conceptual details of the input elements matter by creating a number of instinctively different combinations, while keeping the input elements constant.

The purpose of this experiment was to develop some understanding of the bounds and generality of the computations performed in the LATL regarding exactly what types of representations it combines. Particularly, as numerical quantification did not elicit conceptual combination in the LATL (Del Prato and Pylkkänen, 2014), our focus was on assessing whether this was because the LATL does not perform quantificational operations — which was Del Prato and Pylkkänen's interpretation — or because numbers in general are not a valid input to the LATL's combinatory mechanism. As a critical test case, we employed complex number terms such as *thirty-two*, which at least intuitively, may be instances of conceptual combination with numbers as the input. If such combinations engage the LATL while numerical quantifications do not, this would be evidence that it is the nature of the combinatory operation as opposed to the nature of the input items that matters for the LATL.

Like Del Prato and Pylkkänen (2014), our study employed a production paradigm where subjects named perceptually parallel displays in different ways, depending on task instruction. In all, our design included three combinatory conditions: complex number terms, numerical quantifications, and adjectival modifications, all of which were compared to non-combinatory list controls (Fig. 1). We aimed for minimal lexical differences in the produced utterances, and thus, given that complex number terms involve number words in both first and second position (*thirty two*), we designed the numerical quantifications to also have this property (e.g., *three twos*) while adjectival modifications involved a combination of a color adjective and a number term (*green twos*). As a primary non-combinatory control, we used lists consisting of two single-digit numbers (*two, three*), but also included lists consisting of a decade number and a single-digit number (*thirty, two*), given that lexically, this yields a form identical to the complex number term. However, given that decade numbers are themselves potentially complex, this latter control was not obviously non-combinatory, and thus could have been predicted to pattern somewhere between the combinatory conditions and our single-digit list condition. In fact, this is what we observed (Additional Fig. 1). Thus, we decided to treat the straightforward number list (*two, three*) as the main non-combinatory control.

As depicted in Fig. 1, all our stimuli consisted of rows of four numbers, some of which were colored. In the Color Modification condition, participants were asked to name the color and the identity of the colored numbers as utterances such as "green twos". Although the cardinal name is not a prototypical noun, it is thought to act as a noun when placed as the head of the Noun Phrase (Hurford, 1975, Hurford, 1987, Hurford, 2001,Hurford, 2003 and Ionin and Matushansky, 2006). Our aim was to replicate the prior finding that adjectival modification of nouns engages the LATL (Bemis and Pylkkänen, 2011, Bemis and Pylkkänen, 2013, Del Prato and Pylkkänen, 2014 and Westerlund and Pylkkänen, 2014). Crucially, this condition assessed whether a non-canonical open class word such as a number word could ever elicit LATL-relevant conceptual combination.

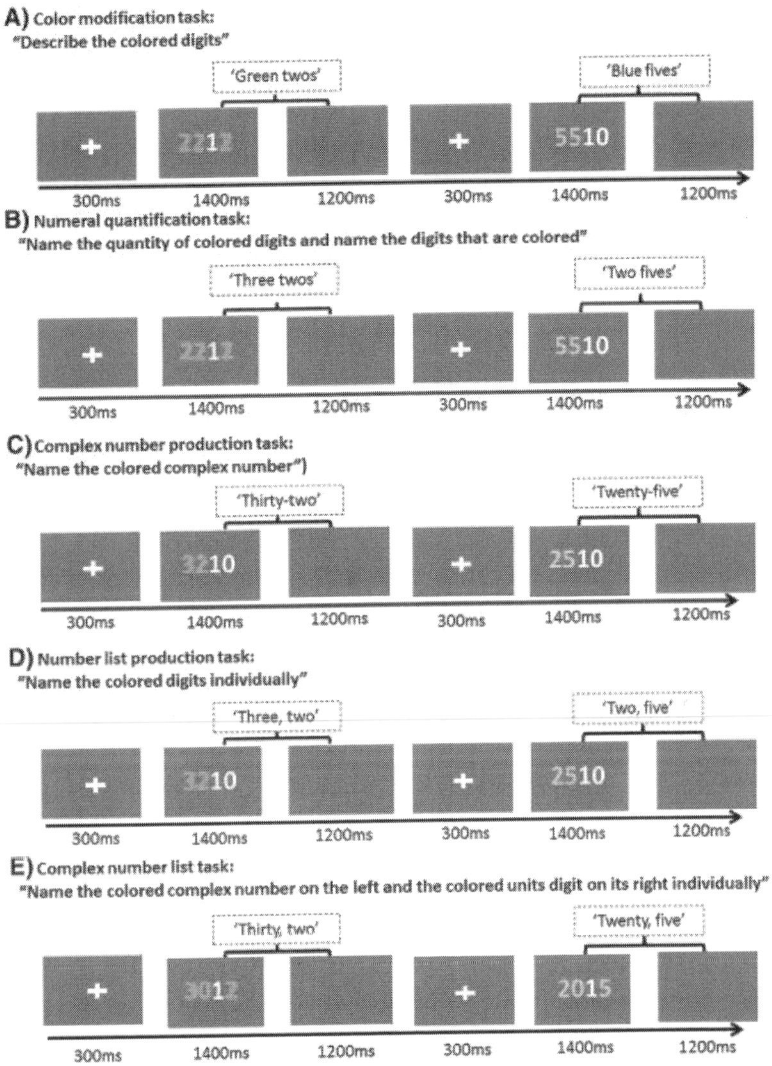

Figure1. Experimental design. Participants were asked to describe the colored digits onscreen by (A) naming the color and the numbers that were colored (B) counting the number of colored digits and naming the digits that were colored, (C) naming the complex number the colored digits formed, (D) naming the complex number on the left and the units number on the right individually or (E) naming the colored digits individually in a list like fashion.

In the Numerical Quantification condition, participants were asked to name aloud the number of colored digits and the name of those digits (e.g. "Three twos"). In this condition, the cardinal number which acts as a modifier characterized the exact cardinality of sets. Thus, in this case, 'three' meant "having exactly three members" (Bale and Khanjian, 2011 and Ionin and Matushansky, 2006). In contrast, in the Complex Number condition, participants were asked to name aloud the two-digit complex number formed by the two colored digits onscreen (e.g., "thirty-two"). Crucially, both this condition and Numerical Quantification combined two number words, but the combinatory operation is different (quantifying the number of numbers vs. creating a complex number). On the basis of Del Prato and Pylkkänen's (2014) findings, we expected no LATL involvement in the processing of the numerical quantifications. If complex numbers patterned similarly, this would suggest that numbers in general cannot function as the "additional feature" whose incorporation constitutes LATL-relevant conceptual combination. In contrast, if complex numbers do engage the LATL, this would indicate that the combinatory operation housed in the LATL does not necessarily require conceptually rich input items, but instead, will also operate on featurally impoverished concepts such as numbers.

So far, we have assumed that at least potentially, the naming of complex numbers involves combinatory operations, but in fact, in the number literature, there has been significant debate about whether two-digit numbers are represented compositionally (i.e., each digit pair is processed as a decade digit and a unit digit separately; e.g. "thirty-two" is the combination of the concept "thirty" and the concept "two") (Grossberg and Repin, 2003; McCloskey, 1992, McCloskey et al., 1986, McCloskey et al., 1985, Nuerk et al., 2001 and Verguts and De Moor, 2005) or holistically (i.e., each digit pair is processed as a single concept: "thirty-two" is stored as a single mental representation) (Brysbaert,

1995,Dehaene et al., 1990, Dotan et al., 2014 and Reynvoet and Brysbaert, 1999). Evidence in favor of holistic representations has included studies reporting that priming effects are constant regardless of whether they involve priming by a single or two-digit number (e.g., the priming effect of 7 on 9 is the same as the priming effect of 11 on 9; Reynvoet and Brysbaert, 1999), studies finding that reaction times for deciding whether a two-digit number is larger or smaller than 65 show no significant discontinuities at decade boundaries (i.e., subjects respond "smaller" more slowly to 59 than to 51, although the 5 in the decades position already indicates that both of these numbers are smaller than 65;Dehaene et al., 1990), and studies showing that number magnitude, frequency of the number and sometimes the syllable length of the number name (but not differences in decade) influenced number reading times (Brysbaert, 1995). The main argument in favor of decomposition has, however, been that it is easier to judge whether a complex number is smaller than another when both the units and the decade digit are smaller (e.g. 67 and 52) than when only one of them is (e.g. 62 and 47) even if controlled for overall numerical distance, which suggests that subjects pay attention to the value of each digit, not the whole number (Nuerk et al., 2001; for more recent evidence see Verguts and De Moor, 2005; for a review see Grossberg and Repin, 2003). Additionally Zhou et al. (2008)suggest that whether complex numbers are processed holistically or compositionally depends on the stage of processing. In sum, then, whether complex numbers involve combinatory processing to begin with is an unsettled empirical question, and one that our dataset should be able to shed novel light on. It should be mentioned that even if the holistic vs decomposed dispute emphases on magnitude judgment, the "triple-code model" assumes that in addition to modality-specific symbolic codes in the visual Arabic and auditory verbal domain, there is also a supramodal abstract "number sense" that conveys semantic information (Dehaene, 1992). Thus, even if the task itself does not explicitly require processing of numerical magnitude this abstract "number sense" remains activated.

A final noteworthy aspect of our design was that both our Color Modification and Numerical quantification conditions involved the production of plural noun phrases (i.e., "blue threes" and "two threes"), whereas the complex number condition did not ("twenty-three"). Previous studies have found increased activation in the left angular gyrus when contrasting plural nouns that were morphosyntactically marked for the number feature (-s) with singular nouns (Domahs et al., 2012) and when linking back two-sentence discourses to plural rather than singular subjects (Boiteau et al., 2014). Given that the left angular gyrus (AG) has also been identified as the most likely candidate for number case agreement violations (Carreiras et al., 2010) and proposed to support the manipulation of numbers in verbal form together with other left perisylvian areas (Dehaene et al., 2003), we analyzed whether the angular gyrus would show effects of morphosyntactic plurality. For completeness, in addition to regions where LATL combinatory effects have localized in prior studies (Brodmann areas 38, 20 and 21), our analysis also included the ventromedial prefrontal cortex, a common though somewhat less consistent locus of basic combinatory effects (e.g., Bemis and Pylkkänen, 2011 and Brennan and Pylkkänen, 2012) as well as the left inferior frontal gyrus (LIFG), given its traditional association with language production, though not, at least in MEG, basic combinatory processing either in comprehension (Bemis and Pylkkänen, 2011 and Bemis and Pylkkänen, 2012) or production (Pylkkänen et al., 2014).

2. MATERIALS AND METHODS

2.1 Participants

28 right-handed, native English speakers participated in this experiment (19 female 9 male, 24.7 years average 4.15 SD). All were neurologically intact, with normal or corrected-to-normal vision and all provided informed written consent.

2.2 Stimuli and experimental design

The experiment consisted of 480 trials in which participants were presented with four digits in a row. The numbers were presented on a dark gray background and some of the digits were colored in pink, blue or green while the rest were presented in light gray (e.g., participants would see "3218" where "3" and "2" were green and "1" and "8" were light gray). The stimuli were kept constant across all conditions, thus assuring that there was no perceptual variation amongst them. Participants were required to name the numbers colored in pink, blue or green aloud; the specifics of the naming task varying upon instruction. Participants were asked to either name the colored digits individually ('three, two'), the quantity of colored digits and the digits that were colored ('three twos'), the color of the digit and the digits that were colored ('green twos'), the complex number that they formed ('thirty-two') or the complex number and the units separately ('thirty, two') (Fig. 1). The quantity of colored digits and their location on the number varied in a controlled fashion between conditions and all numbers were composed by four digits to assure that participants could perceive all the digits of each number at one glance (Kaufman et al., 1949 and Saltzman and Gamer, 1948).

In order to assure that participants were performing a genuine quantification task in the Numerical Quantification condition, we aimed for the base number word (which corresponded to the described colored number) to be bimorphemic (one morpheme in root, and -s). As a consequence, 'one' was excluded as the first-position word and only 'two', 'three' or 'four' were included. The color words for the color modification condition were chosen such as to match the quantifier number words as closely as possible while also being maximally visually distinctive from each other (if the latter constraint was not met, then the condition involving color naming could have turned out harder than number naming). Consequently, we required the color words to be monosyllabic like the number words. For the rest of lexical-level variables, we chose English Lexicon Project naming times as a summary statistic (Balota et al., 2007). The color words 'blue', 'pink' and 'green' were chosen as optimally matching these number words. The English Lexicon Project naming times of these color words were somewhat faster (mean = 577 ms) than those of the Number words (mean = 605 ms), but this difference was not significant ($p = .13$). In addition, only monosyllabic number words from 1 to 9 were included in the design as base numbers. We excluded disyllabic number words (e.g. "seven") to avoid effects related to larger number of syllables (such as delayed naming latencies and greater motor preparation for more syllabled words). Therefore, only eight numbers were used as base numbers ('one', 'two', 'three', 'four', 'five', 'six', 'eight', and 'nine').

All base number words were combined with the three possible first position words (either number words or colors) eliciting 24 combinations for each condition which formed each of the experimental blocks. Each block was repeated four times during the experiment, eliciting 96 trials per condition (480 trials in total). All pictures were presented foveally using Presentation (Neurobehavioral System Inc., California, USA) and subtended in a range from 55.16° height and 33.36° width on a screen ~ 85 cm from the subject.

2.3 Procedure

Before recording, each subject's head shape was digitized using a Polhemus dual source handheld FastSCAN laser scanner (Polhemus, VT, USA). MEG data were collected in the Neuroscience of Language Lab in NYU Abu Dhabi using a whole-head 208 channel axial gradiometer system (Kanazawa Institute of Technology, Kanazawa, Japan) as subjects lay in a dimly lit, magnetically shielded room. Vocal responses were captured with an MEG compatible microphone (Shure PG 81, Shure Europe GmbH).

In all conditions, trials began with a fixation cross (300 ms), followed by the presentation of the stimuli. The picture remained onscreen until speech onset (1400 ms timeout), and participants were allowed 1200 ms to finish their speech before the fixation cross for the following trial would appear. The entire recording lasted ~ 25 min.

3. DATA ACQUISITION AND PREPROCESSING

MEG data were recorded at 1000 Hz (200 Hz low-pass filter), epoched from 200 ms before to 700 ms after picture onset and noise reduced via the Continuously Adjusted Least-Squares Method (Adachi et al., 2001), in the MEG Laboratory software (Yokogawa Electric Corporation and Eagle Technology Corporation, Tokyo, Japan). All trials exceeding the absolute threshold of 2500 fT/cm in amplitude after noise reduction were rejected. Trials containing any remaining blinks were identified by individually visualizing raw activity for each epoch. If there was any sudden, stark increase of activity, the topography for that epoch was plotted. If the magnetic field pattern had the characteristic frontal distribution of a blink, that trial was also rejected. Additionally, trials corresponding to behavioral errors or response times within the length of our epochs were also excluded from further analyses. This procedure resulted in the exclusion of more than 85% of the trials for three participants due to excessive artifacts in their recordings (caused by construction work next door at the time of the experiment). Therefore, these three participants were excluded from further analyses. For the participants included in the analysis, the artifact and blink rejection routines resulted in the exclusion of 27.92% of the trials (11.9% SD), leaving 346.07 trials on average per subject (57.41 SD).

Data were averaged for each condition and subject. Averages were low-pass filtered at 40 Hz. Due to the excellent noise-conditions of the MEG facility, no high pass filtering was required. However, in the course of data analysis, it was observed that baseline correcting with the pre-stimulus interval created an artifactual sustained activity in the MEG averages starting at around 300 ms, due to the fact that the amplitude of the baseline period was higher than the evoked activity after early visual responses. The high baseline amplitude could have been due to task preparation, given that we had a blocked design where subjects needed to ready themselves to name the displays in the appropriate manner given the current task instruction. The drift introduced this way did not however qualitatively alter the obtained results: The same basic pattern was observed whether or not the data were baseline corrected with the pre-stimulus interval. Thus, to avoid visualizing the artifact, in the results reported below, the pre-stimulus interval was not used for baseline correction. Instead, for the purpose of defining the noise covariance matrix used in source analysis, we used an alternative approach offered by BESA Research 6.0, where baseline activity is calculated extracting the values of the 15% of all time points within our epochs that had the lowest global field power.[1]

To estimate the distributed electrical current image in the brain at each time sample we used the Minimum Norm Approach (Hämäläinen & Ilmoniemi, 1994) as implemented in BESA Research 6.0. The sources were evenly distributed using 1500 standard locations 10% and 30% below the smoothed standard brain surface

(750 for each shell). The inverse solution problem was stabilized by the minimum norm mathematical constraint: Out of the many current distributions that could account for the recorded sensor data, the solution with the minimum L2 norm (i.e., the minimum total power of the current distribution) was used. This is performed as follows: First, the forward solution (leadfield matrix L) of all sources was calculated in the current head model. Then, the source activities S(t) of all source components were computed from the data matrix D(t) using an inverse regularized by the estimated noise covariance matrix $[S(t) = R \cdot L^T \cdot (L \cdot R \cdot L^T + C_N)^{-1} \cdot D(t)]$ where L is the leadfield matrix of the distributed regional source model, C_N denotes the noise correlation matrix in sensor space, and R is a weighting matrix in source space. The total activity of each regional source is computed as the root mean square of the source activities S(t) of its 3 (MEG:2) components. Additionally, we applied depth weighting and spatio-temporal weighting. Depth waiting was used in order for both deep and superficial sources to produce a similar, more focal result (as opposed to deep sources appearing very smeared in a minimum-norm reconstruction). This was computed by scaling the leadfield of each regional source with the largest singular value of the SVD (singular value decomposition) of the source's leadfield. The spatio-temporal weighting was conducted to assign large weight to the sources that are assumed to be more likely to contribute to the recorded data. Wc first divided the signal into a signal and a noise subspace. The correlation of the leadfield of a regional source i with the signal subspace (p_i) was computed to find out if the source location contributes to the measured data. The weighting matrix R then becomes a diagonal matrix. Each of the three (MEG: 2) components of a regional source get the same weighting value p_i^2 (Mosher and Leahy, 1998). There was no constraint posited on the dipole orientation (we used free orientation), the regularization constant was 1% and we did not apply any normalization (although we did use the residual variance fit criterion).[2] Regions of interest were defined in terms of

Brodmann areas (BAs), which were isolated with the Talairach Daemon (Lancaster et al., 1997 and Lancaster et al., 2000) from the BESA source space.

4. STATISTICAL ANALYSIS

As the main goal of the current study was to compare combinatory effects in color modification as opposed to either complex number combinations or number quantification, we conducted a main analysis in the areas of the LATL (BAs 38, 20 and 21) that have previously been implicated in conceptual combination (Bemis and Pylkkänen, 2011, Pylkkänen et al., 2014 and Westerlund and Pylkkänen, 2014). Although BA 20 and 21 stretch to more posterior regions of the temporal lobe, they were included in this analysis in order to cover anterior temporal cortex outside of the temporal pole (i.e., BA 38). In prior MEG studies, LATL combinatory effects have centered both around the pole (e.g., Del Prato and Pylkkänen, 2014 and Pylkkänen et al., 2014) as well as more laterally (Westerlund and Pylkkänen, 2014 and Westerlund et al., 2015). Crucially, we complemented our ROI-analyses with liberally thresholded whole brain contrasts capable of revealing the centers of activity within the ROIs.

In addition to the hypothesis-driven LATL analysis, we ran a separate more explorative analysis in areas that do not constitute the main focus of the study but could be sensitive to the current experimental manipulations. This second analysis included the angular gyrus (AG), the ventromedial prefrontal cortex (vmPFC) and the left inferior frontal gyrus (LIFG). The AG was included since we wanted to identify a possible locus for numerical quantification and plural composition, and this area has been previously reported to be involved both in number word processing (Dehaene et al., 2003) and plural representations (Boiteau et al., 2014, Carreiras et al.,

2010 and Domahs et al., 2012). The vmPFC was included as previous studies (Bemis and Pylkkänen, 2011, Bemis and Pylkkänen, 2013 and Pylkkänen et al., 2014) have found the vmPFC to be involved to some extent in basic composition. Following such studies, left and right BA11 were collapsed into a single ROI due to spatial adjacency along the midline. Lastly, although previous MEG studies on language production have not found composition effects in the LIFG (Del Prato and Pylkkänen, 2014 and Pylkkänen et al., 2014) we also included this area in the analysis; given its general prominence in research in production (e.g., Haller et al., 2005, Indefrey et al., 2001a and Menenti et al., 2011). Due to the small number of regional sources within BA 44–45, they were also collapsed into a single ROI. BA 39 was used for the angular gyrus.

For the time-course data of each region, a non-parametric cluster permutation test (Maris and Oostenveld, 2007) with 10,000 permutations was used to identify temporal clusters during which the localized activity differed significantly between conditions, corrected for multiple comparisons over time. Prior to this test, MEG activity was averaged over all sources within a ROI. For initial cluster selection, we adopted the parameters of prior studies: 10 adjacent time points showing an effect at an alpha level of $p < 0.3$, (e.g.,Bemis and Pylkkänen, 2011, Bemis and Pylkkänen, 2012, Bemis and Pylkkänen, 2013,Del Prato and Pylkkänen, 2014, Leiken and Pylkkänen, 2013, Pylkkänen et al., 2014 and Westerlund and Pylkkänen, 2014). Then, for each cluster surviving these thresholds, a test statistic was constructed that was equal to the summed t-values of the point-by-point test-statistics over the selected cluster interval and finally, the cluster with the largest summed test statistic was chosen for further computations. Due to the last step, this test is only capable of identifying one effect within any given analysis interval and thus in order to be able to characterize potential earlier and later effects, all analyses were conducted both in an early (150–400 ms) and a late (400–600 ms) time window. Since all trials with faster reaction times than 700 ms were excluded from the analyses, we ensured these windows did not capture any late motion artifacts. For the largest cluster within an interval the corrected p-value ($p < .05$) was calculated as the ratio of permutations yielding a test statistic greater than the actual observed test statistic. Since only increases for combinatory conditions over lists were interpretable in light of our hypotheses and since our study in general was based on prior evidence that combinatory conditions elicit stronger activity than list conditions (Del Prato and Pylkkänen, 2014 and Pylkkänen et al., 2014), all permutation t-tests were one-tailed. Finally, to moderately protect our analysis against false positives across multiple regions within the same analysis while still maintaining power, a false discovery rate (FDR; Benjamini and Hochberg, 1995 and Genovese et al., 2002) of 0.1 was used throughout. This rate was kept somewhat liberal given that our main research question had no explorative component and the combinatory effects in our main dependent measure, LATL amplitude as measured by MEG, have already been replicated in numerous studies (Bemis and Pylkkänen, 2011, Bemis and Pylkkänen, 2012, Bemis and Pylkkänen, 2013, Westerlund and Pylkkänen, 2014, Del Prato and Pylkkänen, 2014, Pylkkänen et al., 2014, Leffel et al., 2014, Westerlund et al., 2015 and Zhang and Pylkkänen, 2015).

Given that two of our LATL ROIs, BA 20 and 21, covered not only anterior but also posterior temporal cortex, as a final step in our analysis we visualized the centers of any obtained LATL effects with liberally thresholded uncorrected full brain analyses focused on the statistical peaks of the ROI effects in each comparison. Full brain contrasts were run at the time points of the highest uncorrected statistic in the ROI cluster and in the visualization, individual sources were plotted as red (indicating an increase for a combinatory condition) or blue (indicating a decrease for a combinatory condition) when they and at least two of their adjacent spatial and temporal neighbors showed an uncorrected significance of $p < .05$.

As emphasized above, the purpose of these analyses was simply to address any potential spatial ambiguity in the ROI analyses.

5. RESULTS

5.1 Behavioral results

Reaction times were submitted to a one way ANOVA with five levels. The results showed a main effect of condition [$F(4,96) = 84.75$, $p < .0001$]. Participants were the slowest naming the numbers in Color modification condition ($M = 940$ ms; $SD = 210$ ms) and planned *t*-tests showed that this delay was significant when compared to the other two experimental conditions: Numerical quantification [$t(24) = 3.22$, $p = .003$] and Complex Number naming [$t(24) = 10.5$, $p < .0001$]. Additionally, Numerical Quantification was also significantly slower than Complex Number naming [$t(24) = 12.83$, $p < .0001$]. There was no significant difference between our two control conditions; complex number list and number list [$t(24) = 0.83$, $p = .41$], although number list was slightly slower on average ($M = 788$ ms; $SD = 156$ ms for number list vs $M = 778$ ms $SD = 168$ ms for complex number list) (Fig. 2). Accuracy in all conditions was at ceiling, with each participant making an average of 6.3 errors in the course of the whole experiment (0.013%).

5.2 MEG results

In order to find the best control condition for our combinatorial effects, we ran a one way ANOVA with three levels (Complex number, Complex number list and Number list) in the LATL. Although non-significant, the result pattern showed a clear layered effect of composition, with Complex number eliciting the greatest activity followed by Complex number list, and Number list being the one eliciting the least activity (Additional Fig. 1). This pattern suggested that Complex number list condition may have elicited combinatorial activity to some extent. For this reason, we used Number list condition as the baseline control condition to assess combinatorial effects in our experimental manipulations.

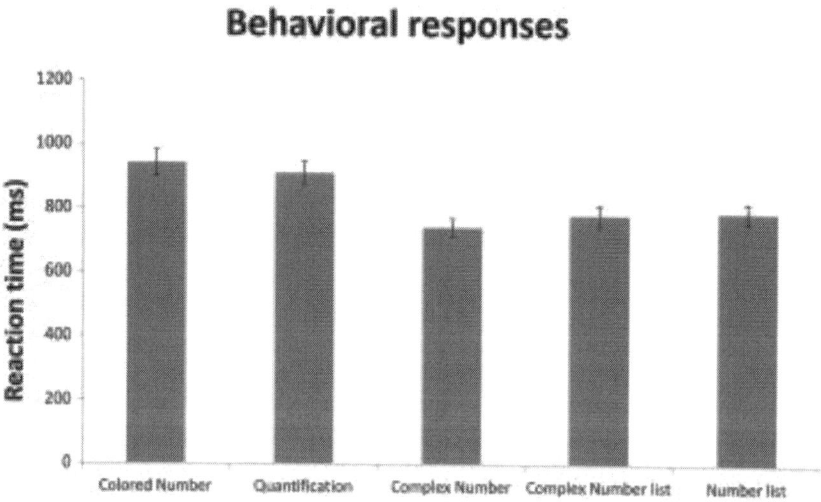

Figure 2. Mean reaction times as a function of the performed naming task. Error bars show standard errors of the mean.

The results of the pair wise comparisons conducted in the LATL (BA 38, 20, 21) revealed increased activation for Color modification over Number lists in all the analyzed areas (Fig. 3). The effect was reliable in BA20 [150–258 ms; p = .04] and in BA38 [184–349 ms;p = .03]. Additionally, a cluster of activity was also identified in BA21, although it did not reach significance [187–244 ms; p = .18]. Complex number condition also elicited significantly greater activity than

Number lists in the three areas. Specifically, this increase in activity was reliable in BA21 [435–589 ms; $p = .02$], and marginally reliable in BA20 [505–590 ms; $p = .07$] and BA38 [513–583 ms; $p = .08$]. In contrast, the comparison between Numerical Quantification and Number List did not elicit significant differences in any of the analyzed areas (no clusters of activation were found in BA21 and BA38, and the cluster located in BA20 did not reach reliability [283–371 ms; $p = .1$]). To further contrast the differences between Complex number and Numerical Quantification, we ran an additional direct comparison between them and the results showed that Complex number elicited reliable increases in activity in BA38 [434–600 ms; $p = .003$], BA20 [437–600 ms; $p = .01$] and BA21 [400–600 ms; $p = .01$] (Fig. 3). Lastly, we also compared Color Modification and Complex number composition directly. These tests did not reveal any reliable differences between these conditions in either the earlier or the later time window.

Thus both complex number formation and color modification modulated the LATL, while numerical quantification did not affect it. When the center of this effect was visualized in the full brain contrasts, it localized ventrally in the anterior temporal cortex for color modification, while the center of the effect was a bit more superior for complex number composition (Fig. 3). Crucially though, both of these localizations are within the range of previously reported effect centers in prior LATL composition studies (Del Prato and Pylkkänen, 2014 and Westerlund and Pylkkänen, 2014). Thus, there is not yet any detailed characterization as regards the precise locus of combinatory effects within the LATL, and given the somewhat fuzzy spatial resolution of MEG, it is also unclear whether such generalizations could be obtained with MEG.

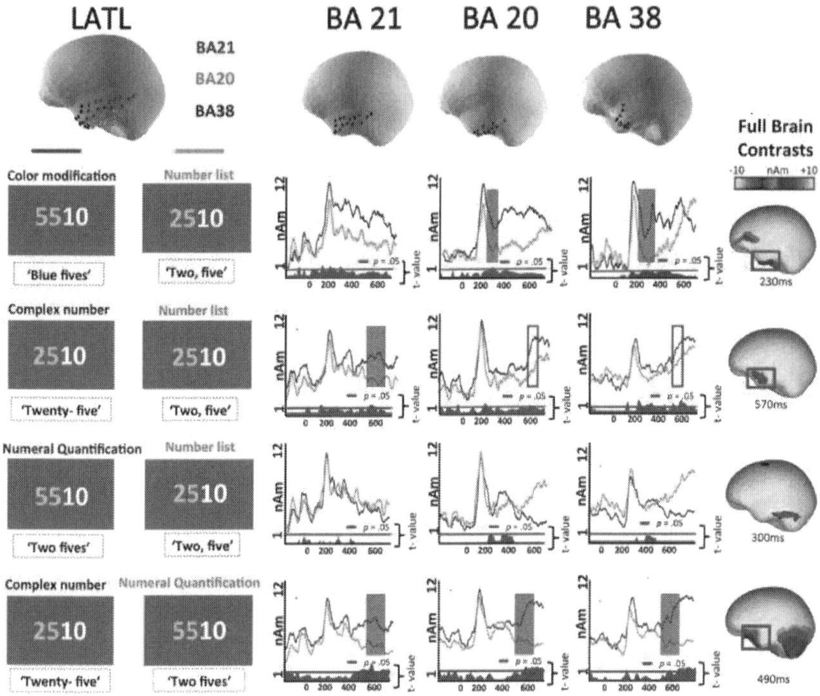

Figure 3. ROI results for pair wise comparisons in the LATL, activation averaged across subjects. On the waveform plots, the shaded regions indicate that the difference in activity between the two tested conditions was significant at a $p = .05$ value (corrected), while the boxed region indicates marginally significant effects ($p < .1$). Significance was determined using a non-parametric, permutation test (Maris and Oostenveld, 2007) performed from 150 to 400 and 400 to 600 ms (10,000 permutations). The point-by-point t-statistic is also plotted in gray, with a red line indicating uncorrected significance at the $p < .05$ level. Finally, the right most panel visualizes the activity centers of the LATL effects obtained in the ROI analysis by plotting a source-by-source full brain comparison at the time of the statistical peak of the ROI effect. Individual sources are plotted as red when they and at least two of their adjacent spatial and temporal neighbors showed an increase for the combinatory condition at $p < .05$.

The analyses on the AG aimed to test sensitivity for plurality in this area. Therefore, conditions containing plural NPs (Color modification and Numerical quantification) and singular NPs (Complex number and Number list) were contrasted. The results did not show significant differences for any of these contrasts. The vmPFC was reliably engaged when composing a complex number in comparison to naming a number list [458–600 ms;p = .01] but not when performing a numerical quantification (we did not find any cluster of activity) or performing a color modification [186–223 ms; p = .22]. Lastly, no comparisons elicited reliable clusters of activation in the LIFG (Additional Fig. 2).

6. DISCUSSION

6.1 Complex number composition in the LATL

In this study we aimed to develop some understanding into the specifics of the combinatorial processes performed in the LATL and the nature of the elements that can enter these computations. Particularly, we tried to elucidate if the lack of LATL engagement during numerally quantified phrases such as *two cups* in previous production studies (Del Prato and Pylkkänen, 2014) was because number words are not a valid input element for the LATL or because the LATL is not involved in the computations underlying quantification in particular. To achieve this, we tested complex number expressions (*thirty-three*) where the combinatory mode intuitively resembles conceptual combination, already demonstrated to drive the LATL (comprehension: Baron and Osherson, 2011, Bemis and Pylkkänen, 2011, Bemis and Pylkkänen, 2013 and Westerlund and Pylkkänen, 2014; production: Del Prato and Pylkkänen, 2014 and Pylkkänen et al., 2014), and compared this to numerical quantification. According to our results, the reason for the lack of LATL effects in numerical quantifications is not the numerical input but rather the mode of composition: when numbers compose into complex concepts as in *thirty-two*, LATL activity is increased in comparison to both non-combinatory lists (*three, two*) as well as to quantificational phrases (*three twos*). Thus our findings suggest that the LATL is not a general purpose combiner of meanings but rather specializes in some version of conceptual combination, potentially delimited to situations where one combining element characterizes a property of the other.

Additionally, the finding of combinatorial activity for our complex number condition conforms to theories where complex numbers undergo a composition process before being produced (Deloche and Seron, 1987), as opposed to being processed holistically (Brysbaert, 1995, Dehaene et al., 1990 and Reynvoet and Brysbaert, 1999). In other words, our results suggest that in order to produce "twenty-three", the concept of "twenty" and the concept of "three" are retrieved and combined. This proposal is consistent withMcCloskey et al.'s (1986) model where complex number production involves the generation of a syntactic frame that specifies each to-be-retrieved word in terms of a number-lexical class (ones, teens...) and a position within that class. Moreover, it conforms to proposals within theoretical linguistics where complex number composition follows the standard principles of semantic and syntactic combination (Ionin and Matushansky, 2006).

Despite the anatomical overlap in regions activated by complex number naming and color modification, we did observe an unpredicted difference in the time course of these effects: Composition effects for complex numbers occurred around 200 ms later than the adjectival modification effects. This difference is especially interesting given that articulation onsets exhibited the reverse pattern: complex number productions were faster than color modifications (though the plural morphology of the color modification likely contributed to this; see next section). This makes explanations in terms of any type of general difficulty somewhat unlikely: for example, the two conditions differed in the number of syllables, but under a syllable-based account, it would be unclear how faster reaction times could be elicited when more syllables are articulated. Instead, inspection of the source waveforms (Fig. 3) suggests that in general, similar peaks and valleys are elicited both for color modification and complex number formation, but along this waveform morphology, the two conditions show significant effects at different stages. Given that language production is thought to involve a complex interplay of both comprehension and production processes (e.g., Pickering and Garrod, 2013), one could then speculate that the combinatory processing of complex numbers occurs most strongly at a later stage in this sequence than the combinatory processing of adjectival modifications. Thus in future work it will be possible to articulate and test more specific hypotheses regarding such a contrast.

7. CONTRIBUTIONS OF THE AG, VMPFC AND LIFG

The AG was included in our analysis to test whether we would replicate the finding byDomahs et al. (2012) that plural nouns elicit reliably greater AG activity than singular nouns. We tested this by comparing the two conditions that involved naming a plural NP (Color modification and Numerical Quantification) to the two that involved a singular NP (Complex number and Number list), but did not find any reliable differences between conditions. Interestingly, Domahs et al.'s study did not find the AG to be uniquely involved in the processing of plural entities per se, but rather in the processing of non-singular items (including mass nouns). Thus, a possible reason for the lack of differences between our conditions could be that all of our experimental conditions, including the singular NPs, were composed of number words which represented non-singular concepts. This interpretation of the AG as a non-singularity representation hub but not a syntactic number marking region is also convergent with Carreiras et al. (2010), who did not find the AG to be involved in number agreement processing.

Interestingly, in the behavioral data, we observed longer naming latencies in exactly the two conditions that involved naming a plural NP (Color modification and Numerical Quantification) as compared to the three conditions that involved naming singular NPs. This "plural effect" conforms to prior findings in comprehension showing delayed responses to plural as compared to singular NPs (Tucker et al., 2015, Lau et al., 2008 and Wagers et al., 2009), suggesting that the processing of syntactically plural entities is more effortful than singular ones. However, our results do not speak to the neural substrates of this phenomenon; thus, future work will need to unveil the correlates of syntactic number marking and will have to establish the relationship between grammatical number case and general number magnitude processing.

The analyses of vmPFC activity revealed trends towards increases for complex numbers and color modifications over number list whereas quantification did not elicit higher activity than the list. These results provide moderate support for an implication of the vmPFC in basic combination (Bemis and Pylkkänen, 2011 and Pylkkänen et al., 2014) and suggest that numerical input can also drive combinatorial activity in the vmPFC.

Finally, consistent with prior MEG studies on basic combinatory phrases both in comprehension (Bemis and Pylkkänen, 2011 and Bemis and Pylkkänen, 2012) and production (Del Prato and Pylkkänen, 2014 and Pylkkänen et al., 2014), we did not observe combinatory effects in the LIFG. Thus to the extent that the LIFG may be involved in linguistic composition (e.g., Hagoort, 2005), the nature of its contribution is not a time-locked evoked response of the sort that would be revealed by the type of source modeling of averaged data as performed here.

8. CONCLUSION

In this work we set out to delineate the combinatory computation housed in the LATL, with a focus on number words and their composition. We found that the LATL composes number words but only if the composing items combine into a more complex concept (*thirty two*) and not when one word enumerates the number of tokens of the other (*three twos*). Our findings suggest that the engagement of the LATL is determined by the computations underlying the performed combinatorial process as opposed to the nature of the input items.

ACKNOWLEDGMENTS

This research was supported by the National Science Foundation grant BCS-1221723(LP) and grant G1001 from the NYUAD Institute, New York University Abu Dhabi (LP).

REFERENCES

1. Adachi, Y., Shimogawara, M., Higuchi, M., Haruta, Y., Ochiai, M., 2001. Reduction of non- periodic environmental magnetic noise in MEG measurement by continuously ad- justed least squares method. IEEE Trans. Appl. Supercond. 11, 669–672.

2. Alario, F.X., Costa, A., Caramazza, A., 2002. Frequency effects in noun phrase production: implications for models of lexical access. Lang. Cogn. Process. 17 (3), 299–319.

3. Bale, A.C., Khanjian, H., 2011. Cross-Linguistic Representations of Numerals and Number Marking. Proceedings of SALT vol. 20, pp. 582–598.

4. Balota, D.A., Yap, M.J., Hutchison, K.A., Cortese, M.J., Kessler, B., Loftis, B., Neely, J.H., Nelson, D.L., Simpson, G.B., Treiman, R., 2007. The English Lexicon Project. Behav. Res. Methods 39 (3), 445–459.

5. Baron, S.G., Osherson, D., 2011. Evidence for conceptual combination in the left anterior temporal lobe. NeuroImage 55 (4), 1847–1852.

6. Barwise, J., Cooper, R., 1981. Generalized quantifiers and natural language. Linguist. philos. 4 (2), 159–219.

7. Bemis, D.K., Pylkkänen, L., 2011. Simple composition: a magnetoencephalography inves- tigation into the comprehension of minimal linguistic phrases. J. Neurosci. 31 (8), 2801.

8. Bemis, D.K., Pylkkänen, L., 2012. Basic linguistic composition recruits the left anterior tem- poral lobe and left angular gyrus during both listening and reading. Cereb. Cortex 23 (8), 1859–1873.

9. Bemis, D.K., Pylkkänen, L., 2013. Basic linguistic composition recruits the left anterior tem- poral lobe and left angular gyrus during both listening and reading. Cereb. Cortex 23 (8), 1859–1873.

10. Benjamini, Y., Hochberg, Y., 1995. Controlling the false discovery rate: a practical and powerful approach to multiple testing. J. R. Stat. Soc. Ser. B Methodol. 289–300.

11. Bennett, M.R., 1975. Some Extensions of a Montague Fragment of English. vol. 129. Indiana University Linguistics Club.

12. Binder, J.R., Desai, R.H., 2011. The neurobiology of semantic memory. Trends Cogn. Sci. 15 (11), 527–536.

13. Binder, J.R., Desai, R.H., Graves, W.W., Conant, L.L., 2009. Where is the semantic system? A critical review and meta-analysis of 120 functional neuroimaging studies. Cereb. Cortex 19 (12), 2767–2796.

14. Boiteau, T.W., Bowers, E., Nair, V.A., Almor, A., 2014. The neural representation of plural discourse entities. Brain Lang. 137, 130–141.

15. Bonner, M.F., Peelle, J.E., Cook, P.A., Grossman, M., 2013. Heteromodal conceptual process- ing in the angular gyrus. NeuroImage 71, 175–186.

16. Brennan, J., Pylkkänen, L., 2012. The time-course and spatial distribution of brain activity associated with sentence processing. NeuroImage 60 (2), 1139–1148.

17. Bright, P., Moss, H., Tyler, L., 2004. Unitary vs multiple semantics: PET studies of word and picture processing. Brain Lang. 89 (3), 417–432.

18. Brysbaert, M., 1995. Arabic number reading: on the nature of the numerical scale and the origin of phonological recoding. J. Exp. Psychol. Gen. 124, 434–452.

19. Carreiras, M., Carr, L., Barber, H.A., Hernandez, A., 2010. Where syntax meets math: right intraparietal sulcus activation in response to grammatical number agreement viola- tions. NeuroImage 49 (2), 1741–1749.

20. Clarke, A., Taylor, K.I., Tyler, L.K., 2011. The evolution of meaning: spatio-temporal dynam- ics of visual object recognition. J. Cogn. Neurosci. 23 (8), 1887–1899.

21. Clarke, A., Taylor, K.I., Devereux, B., Randall, B., Tyler, L.K., 2013. From perception to concep- tion: how meaningful objects are processed over time. Cereb. Cortex 23 (1), 187–197.

22. Dehaene, S., 1992. Varieties of numerical abilities. Cognition 44 (1), 1–42.

23. Dehaene, S., Dupoux, E., Mehler, J., 1990. Is numerical comparison digital? Analogical and symbolic effects in two-digit number comparison. J. Exp. Psychol. Hum. Percept. Per- form. 16, 626–641.

24. Dehaene, S., Piazza, M., Pinel, P., Cohen, L., 2003. Three parietal circuits for number pro- cessing. Cogn. Neuropsychol. 20 (3–6), 487–506.

25. Del Prato, P., Pylkkänen, L., 2014. MEG evidence for conceptual combination but not nu- meral quantification in the left anterior temporal lobe during language production. Front. Psychol. 5 (524).

26. Deloche, G., Seron, X., 1987. Numerical Transcoding: A General Production Model. Lawrence Erlbaum Associates, Inc.

27. Domahs, F., Nagels, A., Domahs, U., Whitney, C., Wiese, R., Kircher, T., 2012. Where the mass counts: common cortical activation for different kinds of nonsingularity. J. Cogn. Neurosci. 24 (4), 915–932.

28. Dotan, D., Friedmann, N., Dehaene, S., 2014. Breaking down number syntax: spared com- prehension of multi-digit numbers in a patient with impaired digit-to-word conver- sion. Cortex 59, 62–73.

29. Friederici, A.D., 2011. The brain basis of language processing: from structure to function. Physiol. Rev. 91 (4), 1357–1392.

30. Friederici, A.D., Meyer, M., Von Cramon, D.Y., 2000. Auditory language comprehension: an event-related fMRI study on the processing of syntactic and lexical information. Brain Lang. 74 (2), 289–300.

31. Gauthier, I., Anderson, A.W., Tarr, M.J., Skudlarski, P., Gore, J.C., 1997. Levels of categoriza- tion in visual recognition studied using functional magnetic resonance imaging. Curr. Biol. 7 (9), 645–651.

32. Genovese, C.R., Lazar, N.A., Nichols, T., 2002. Thresholding of statistical maps in functional neuroimaging using the false discovery rate. NeuroImage 15 (2), 870–878.

33. Grabowski, T.J., Damasio, H., Tranel, D., Ponto, L.L.B., Hichwa, R.D., Damasio, A.R., 2001. A role for left temporal pole in the retrieval of words for unique entities. Hum. Brain Mapp. 13 (4), 199–212.

34. Grossberg, S., Repin, D.V., 2003. A neural model of how the brain represents and compares multi-digit numbers: spatial and categorical processes. Neural Netw. 16 (8), 1107–1140. Hagoort, P., 2005. On Broca, brain, and binding: a new framework. Trends Cogn. Sci. 9 (9), 416–423.

35. Hagoort, P., Indefrey, P., 2014. The neurobiology of language beyond single words. Annu. Rev. Neurosci. 37, 347–362.

36. Haller, S., Radue, E.W., Erb, M., Grodd, W., Kircher, T., 2005. Overt sentence production in event-related fMRI. Neuropsychologia 43, 807–814.

37. Hämäläinen, M.S., Ilmoniemi, R.J., 1994. Interpreting magnetic fields of the brain: mini- mum norm estimates. Med. Biol. Eng. Comput. 32 (1), 35–42.

38. Hampton, J.A., 1997. Conceptual combination. In: Lamberts, K., Shanks, D. (Eds.), Knowledge, Concepts, and Categories. UCL Press, London, pp. 133–159.

39. Heim, S., Eickhoff, S.B., Amunts, K., 2008. Specialisation in Broca's region for semantic, phonological, and syntactic fluency? NeuroImage 40 (3), 1362–1368.

40. Humphries, C., Binder, J.R., Medler, D.A., Liebenthal, E., 2006. Syntactic and semantic mod- ulation of neural activity during auditory sentence comprehension. J. Cogn. Neurosci. 18 (4), 665–679.

41. Humphries, C., Binder, J.R., Medler, D.A., Liebenthal, E., 2007. Time course of semantic pro- cesses during sentence comprehension: an fMRI study. NeuroImage 36 (3), 924–932.

42. Hurford, J., 1975. The Linguistic Theory of Numerals. Cambridge University Press, Cambridge.

43. Hurford, J., 1987. Language and Number: The Emergence of a Cognitive System. Blackwell, Oxford.

44. Hurford, J., 2001. Numeral systems. In: Smelser, N.J., Baltes, P.B. (Eds.), International Encyclopedia of the Social and Behavioral Sciences. Pergamon, Amsterdam, pp. 10756–10761.

45. Hurford, J., 2003. The Interaction Between Numerals and Nouns. In: Plank, F. (Ed.), Noun Phrase Structure in the Languages of Europe. Typology of Languages in Europe 7. Mouton de Gruyter. The Hague.

46. Indefrey, P., 2012. Hemodynamic studies of syntactic processing. In: Faust, M. (Ed.), The Handbook of the Neuropsychology of Language. MA: Blackwell, Malden, pp. 209–228.

47. Indefrey, P., Brown, C.M., Hellwig, F., Amunts, K., Herzog, H., Hagoort, P., 2001a. A neural correlate of syntactic encoding during speech production. Proc. Natl. Acad. Sci. 98, 5933–5936.

48. Indefrey, P., Hagoort, P., Herzog, H., Seitz, R.J., Brown, C.M., 2001b. Syntactic processing in left prefrontal cortex is independent of lexical meaning. NeuroImage 14 (3), 546–555.

49. Ionin, T., Matushansky, O., 2006. The composition of complex cardinals. J. Semant. 23 (4), 315–360.

50. Kaufman, E.L., Lord, M.W., Reese, T.W., Volkmann, J., 1949. The discrimination of visual number. Am. J. Psychol. 62, 498–525.

51. Lancaster, J.L., Rainey, L.H., Summerlin, J.L., Freitas, C.S., Fox, P.T., Evans, A.C., ... Mazziotta, J.C., 1997. Automated labeling of the human brain: a preliminary re- port on the development and evaluation of a forward-transform method. Hum. Brain Mapp. 5 (4), 238.

52. Lancaster, J.L., Woldorff, M.G., Parsons, L.M., Liotti, M., Freitas, C.S., Rainey, L., ... Fox, P.T., 2000. Automated Talairach atlas labels for functional brain mapping. Hum. Brain Mapp. 10 (3), 120–131.

53. Lau, E.F., Phillips, C., Poeppel, D., 2008. A cortical network for semantics:(de) constructing the N400. Nat. Rev. Neurosci. 9 (12), 920–933.

54. Leffel, T., Lauter, M., Westerlund, M., Pylkkänen, L., 2014. Restrictive vs. non-restrictive composition: a magnetoencephalography study. Lang. Cogn. Neurosci. 29 (10), 1191–1204.

55. Leiken, K., Pylkkänen, L., 2013. MEG evidence that the LIFG effect of object extraction re- quires similarity-based interference. Language and Cognitive Processes.

56. Link, G., 2002. The logical analysis of plurals and mass terms: a lattice-theoretical ap- proach. Formal Semantics: The Essential Readingspp. 127–146.

57. Maris, E., Oostenveld, R., 2007. Nonparametric statistical testing of EEG- and MEG-data. J. Neurosci. Methods 164 (1), 177–190.

58. Mazoyer, B.M., Tzourio, N., Frak, V., Syrota, A., Murayama, N., Levrier, O., et al., 1993. The cortical representation of speech. J. Cogn. Neurosci. 5 (4), 467–479.

59. McCloskey, M., 1992. Cognitive mechanisms in numerical processing: evidence from ac- quired dyscalculia. Cognition 44 (1), 107–157.

60. McCloskey, M., Caramazza, A., Basili, A., 1985. Cognitive mechanisms in number process- ing and calculation: evidence from dyscalculia. Brain Cogn. 4 (2), 171–196.

61. McCloskey, M., Sokol, S.M., Goodman, R.A., 1986. Cognitive processes in verbal-number production: Inferences from the performance of brain-damaged subjects. J. Exp. Psychol. Gen. 115 (4), 307.

62. Medin, D.L., Shoben, E.J., 1988. Context and structure in conceptual combination. Cogn. Psychol. 20 (2), 158–190.

63. Menenti, L., Gierhan, S.M.E., Segaert, K., Hagoort, P., 2011. Shared language: overlap and segregation of the neuronal infrastructure for speaking and listening revealed by functional MRI. Psychol. Sci. 22, 1173–1182.

64. Mesulam, M.M., 2000. Behavioral neuroanatomy: large-scale networks, association cor- tex, frontal syndromes, the limbic system, and hemispheric specializations. Princ. Behav. Cogn. Neurol. 2, 1–120.

65. Meyer, A.S., 1996. Lexical access in phrase and sentence production: results from picture– word interference experiments. J. Mem. Lang. 35, 477–496.

66. Montague, R., 1974. Formal Philosophy; Selected Papers of Richard Montague.

67. Mosher, J.C., Leahy, R.M., 1998. Recursive MUSIC: a framework for EEG and MEG source localization. Biomed. Eng.IEEE Trans. 45 (11), 1342–1354.

68. Murphy, G.L., 1990. Noun phrase interpretation and conceptual combination. J. Mem. Lang. 29 (3), 259–288.

69. Nuerk, H.C., Weger, U., Willmes, K., 2001. Decade breaks in the mental number line? Put- ting tens and units back into different bins. Cognition 82, B25–B33.

70. Pallier, C., Devauchelle, A.D., Dehaene, S., 2011. Cortical representation of the constituent structure of sentences. Proc. Natl. Acad. Sci. 108 (6), 2522.

71. Pandya, D.N., Seltzer, B., 1982. Association areas of the cerebral cortex. Trends Neurosci. 5, 386–390.

72. Pickering, M.J., Garrod, S., 2013. An integrated theory of language production and compre- hension. Behav. Brain Sci. 36 (04), 329–347.

73. Price, A.R., Bonner, M.F., Peelle, J.E., Grossman, M., 2015. Converging evidence for the neu- roanatomic basis of combinatorial semantics in the angular gyrus. J. Neurosci. 35 (7), 3276–3284.

74. Pylkkänen, L., Bemis, D.K., Blanco Elorrieta, E., 2014. Building phrases in language produc- tion: an MEG study of simple composition. Cognition 133, 371–384.

75. Reynvoet, B., Brysbaert, M., 1999. Single-digit and two-digit Arabic numbers address the same semantic number line. Cognition 72, 191–201.

76. Rogalsky, C., Hickok, G., 2009. Selective attention to semantic and syntactic features modulates sentence processing networks in anterior temporal cortex. Cereb. Cortex 19 (4), 786.

77. Rogers, T.T., Hocking, J., Noppeney, U., Mechelli, A., Gorno-Tempini, M.L., Patterson, K., et al., 2006. Anterior temporal cortex and semantic memory: reconciling findings from neuropsychology and functional imaging. Cogn. Affect. Behav. Neurosci. 6 (3), 201–213.

78. Saltzman, I.J., Gamer, W.R., 1948. Reaction time as a measure of span of attention. Am. J. Psychol. 25, 227–241.

79. Scha, R., 1984. Distributive, Collective and Cumulative Quantification. Truth, Interpreta- tion, and Information, GRASS 2 pp. 131–158.

80. Schriefers, H., De Ruiter, J.P., Steigerwald, M., 1999. Parallelism in the production of noun phrases: experiments and reaction time models. J. Exp. Psychol. Learn. Mem. Cogn. 25 (3), 702.

81. Segaert, K., Menenti, L., Weber, K., Petersson, K.M., Hagoort, P., 2012. Shared syntax in lan- guage production and language comprehension—an fMRI study. Cereb. Cortex 22 (7), 1662–1670.

82. Stowe, L.A., Broere, C.A.J., Paans, A.M.J., Wijers, A.A., Mulder, G., Vaalburg, W., et al., 1998. Localizing components of a complex task: sentence processing and working memory. NeuroReport 9 (13), 2995.

83. Stromswold, K., Caplan, D., Alpert, N., Rauch, S., 1996. Localization of syntactic compre- hension by positron emission tomography. Brain Lang. 52 (3), 452–473.

84. Thompson-Schill, S.L., D'Esposito, M., Aguirre, G.K., Farah, M.J., 1997. Role of left inferior prefrontal cortex in retrieval of

semantic knowledge: a reevaluation. Proc. Natl. Acad. Sci. U. S. A. 94, 14792–14797.

85. Tucker, M.A., Idrissi, A., Almeida, D., 2015. Representing number in the real-time process- ing of agreement: self-paced reading evidence from Arabic. Front. Psychol. 6, 314.

86. Tyler, L.K., Stamatakis, E.A., Bright, P., Acres, K., Abdallah, S., Rodd, J., et al., 2004. Process- ing objects at different levels of specificity. J. Cogn. Neurosci. 16 (3), 351–362.

87. Tyler, L.K., Marslen-Wilson, W.D., Randall, B., Wright, P., Devereux, B.J., Zhuang, J., ... Stamatakis, E.A., 2011. Left inferior frontal cortex and syntax: function, structure and behaviour in patients with left hemisphere damage. Brain 134 (2), 415–431.

88. Vandenberghe, R., Nobre, A.C., Price, C.J., 2002. The response of left temporal cortex to sentences. J. Cogn. Neurosci. 14 (4), 550–560.

89. Verguts, T., De Moor, W., 2005. Two-digit comparison: decomposed, holistic, or hybrid? Exp. Psychol. 52 (3), 195–200.

90. Wagers, M.W., Lau, E.F., Phillips, C., 2009. Agreement attraction in comprehension: repre- sentations and processes. J. Mem. Lang. 61 (2), 206–237.

91. Westerlund, M., Pylkkänen, L., 2014. The role of the left anterior temporal lobe in seman- tic composition vs. semantic memory. Neuropsychologia 57, 59–70.

92. Westerlund, M., Kastner, I., Al Kaabi, M., Pylkkänen, L., 2015. The LATL as locus of compo- sition: MEG evidence from English and Arabic. Brain Lang. 141, 124–134.

93. Wisniewski, E.J., 1996. Construal and similarity in conceptual combination. J. Mem. Lang. 35 (3), 434–453.

94. Xu, J., Kemeny, S., Park, G., Frattali, C., Braun, A., 2005. Language in context: emergent fea- tures of word, sentence, and narrative comprehension. NeuroImage 25 (3), 1002–1015.

95. Yeterian, E.H., Pandya, D.N., 1985. Corticothalamic connections of the posterior parietal cortex in the rhesus monkey. J. Comp. Neurol. 237 (3), 408–426.

96. Zhang, L., Pylkkänen, L., 2015. The Interplay of Composition and Concept Specificity in the Left Anterior Temporal Lobe: An MEG Study. NeuroImage 111, 228–240.

97. Zhou, X., Chen, C., Chen, L., Dong, Q., 2008. Holistic or compositional representation of two-digit numbers? Evidence from the distance, magnitude, and SNARC effects in a number-matching task. Cognition 106 (3), 1525–1536.

CHAPTER 4

Combinatorial Interpretation of Raney Numbers and Tree Enumerations

Chin Hee Pah[1], Mohamed Ridza Wahiddin[2]

[1]Department of Computational and Theoretical Sciences, Faculty of Science, International Islamic University Malaysia, Kuantan, Malaysia

[2]Department of Computer Science, Faculty of ICT, International Islamic University Malaysia, Kuala Lumpur, Malaysia

ABSTRACT

A new combinatorial interpretation of Raney numbers is proposed. We apply this combinatorial interpretation to solve several tree enumeration counting problems. Further a generalized Catalan triangle is introduced and some of its properties are proved.

KEYWORDS

Raney Numbers, Fuss-Catalan Numbers, Tree Enumeration, Network

1. INTRODUCTION

Interestingly Penson and Zyczkowski were the first who use the term Raney numbers [1] [2] and it is defined as $R_k(n,r)=\frac{r}{kn+r}\binom{kn+r}{n}$ where $k\geq 2, n\geq 1, r\geq 1$. Nevertheless, it is known that Raney's lemma could be used in counting problem associated with Catalan numbers [3] and a bijection exists between Raney path and plan multitree [4] .

These numbers do not form novel sequences, as the numbers were introduced earlier as a generalization of the binomial series [5] . Moreover, the sequence

$$R_4(n,5)=1,5,30,200,1425,10626,81900,647280,\ldots$$

is not included in OEIS database [6] before 2011. If we let $r=1$, we obtain another known sequence, i.e.,

Fuss-Catalan numbers [7] [8] which is defined as $C_k(n)=\frac{1}{(k-1)n+1}\binom{kn}{n}$. Although Fuss-Catalan numbers were introduced earlier than Catalan numbers [9] , the Catalan numbers are more popular and widely used than the Fuss-Catalan numbers (see [10] [11] for details). Due to its self similar structure, the applications of Catalan numbers could be found in many physical problems, e.g., lattice model [12] , tree enumeration network [13] , and Hankel matrices in coding theory [14] . A tree is a connected graph with no cycles and for which only one shortest path exists from one node to another. Tree enumeration is an important tool to study network. These networks always grow in a power-law behavior which is often found in social network, subway system [15] , etc.

In this paper, we introduce Raney numbers $R(k,r)$ in the form of a non-linear recursion and then we pro- vide a combinatorial interpretation of Raney numbers. Using this combinatorial interpretation, we solve several tree enumeration counting problems in which we recover the well-known Fuss-Catalan numbers [16] , Catalan triangles [17] , and other less known numbers. Motivated by the connection between Raney numbers and Catalan triangles, a generalization of Catalan triangles is proposed and we prove some of their properties. Consequently these formulas generalize the properties of Catalan triangles. From the exact solution of these tree enumeration problems, we are able to find a sharp upper bound of the number of each tree enumeration problem. The upper bound is important in the contour method for lattice models and limit of the random graph.

2. RANEY NUMBERS

Let $C_k(n)$ be the number of a k-ary trees with labeled n vertices (Figure 1), where

$$C_k(n)=\frac{1}{(k-1)n+1}\binom{kn}{n}, \quad k\geq 2, n\geq 1.$$

The Raney numbers are defined as follows:

$$R_k(n,r)=\sum_{i_1+i_2+\cdots+i_r=n} C_k(i_1)C_k(i_2)\cdots C_k(i_r),\quad C_k(0)=1,\quad \forall n>0 \tag{1}$$

where $i_1,i_2,\ldots,i_r \in \mathbb{N}\cup\{0\}$. Therefore, the combinatorial interpretation is as follows: r copies of k-ary tree with total number of n vertices.

Next, we let $u_k(x)$ be the generating function for $C_k(n)$, i.e.,

$$u_k(x)=1+C_k(1)x+C_k(2)x^2+\cdots+C_k(n)x^n+\cdots.$$

Then, the generating function of $R_k(n,r)$ is $u_k^r(x)$ and the Raney numbers satisfy the following formula [18] .

Lemma 1. Let $u_k^r(x)$ be the generating function of the Raney numbers. Then,

$$\left[x^n\right]\left(u^r(x)\right)=\frac{r}{kn+r}\binom{kn+r}{n}. \tag{2}$$

Immediately, we obtain the following theorem.

Theorem 1. The binomial forms of the Raney numbers are given by

$$R_k(n,r) = \frac{r}{kn+r}\binom{kn+r}{n}. \tag{3}$$

From theorem (1), it is not difficult to deduce some of the properties of Raney numbers.

Corollary 1. For integer $k > 1$, we have

$$R_k(0,r) = 1; \tag{4}$$

Figure 1. A binary tree with 3 nodes, where the bottom vertex is the root.

$$R_k(n-i,ki) = \frac{i}{n}\binom{kn}{n-i}; \tag{5}$$

$$R_k(n-1,k)=R_k(n,1)=\frac{1}{(k-1)n+1}\binom{kn}{n}; \tag{6}$$

$$R_k(n,n)=\binom{kn+n-1}{n-1}. \tag{7}$$

Corollary 2. We can write $C_k(n)$ in a nonlinear recursion as:

$$C_k(n+1)=\sum_{i_1+i_2+\cdots+i_k=n} C_k(i_1)C_k(i_2)\cdots C_k(i_k),\quad C_k(0)=1 \tag{8}$$

where $i_1,i_2,\cdots,i_k \in \mathbb{N}\cup\{0\}$.

We recover the formula by joining the k copies of k-ary tree with n vertices which is also equivalent to a k-ary tree with n vertices and an additional root (see Figure 2):

$$R_k(n,k)=R_k(n+1,1).$$

Using binomial form of $R_k(n,r)$, one can obtain the following result.

Corollary 3. For a fixed integer $k>1$, and $n\geq r>1$,

$$R_k(n-1, r+k-1) = R_k(n,r) - R_k(n,r-1) \tag{9}$$

where $R_k(n,r)=0$ if $n<r$.

For $k=2$, we recover the identity of a generalized Ballot numbers:

$$R_2(n-1, r+1) = R_2(n,r) - R_2(n,r-1). \tag{10}$$

3. A HOMOGENEOUS K-ARY TREE

Unlike the usual k-ary tree, we define a homogeneous k-ary tree as a graph with no cycles, in which each vertex emanates $k+1$ edges (see Figure 3 for k = 4). We fix a vertex namely z as the root. Unlike the ordinary root in a k-ary tree, this root has $k+1$ successors while other vertices have k number of successors. Any vertex could be chosen to be the root since the graph is homogenous. For a given n vertices, we may find how many connected sub-tree rooted at z. This number is defined as $D_k(n)$.

Theorem 2. For $k>1$, we can write $D_k(n)$ in a nonlinear recursion of $C_k(n)$ as:

$$D_k(n) = \sum_{r=1}^{n} C_k(r) C_k(n-r) \quad \text{for } n > 0. \tag{11}$$

Proof. We decompose the problem by finding out the number of k-ary tree of n number of one copy of k-ary tree with r vertices, i.e., $C_k(r)$, and another copy of k-ary tree with $n-r$ vertices, i.e. $C_k(n-r)$.

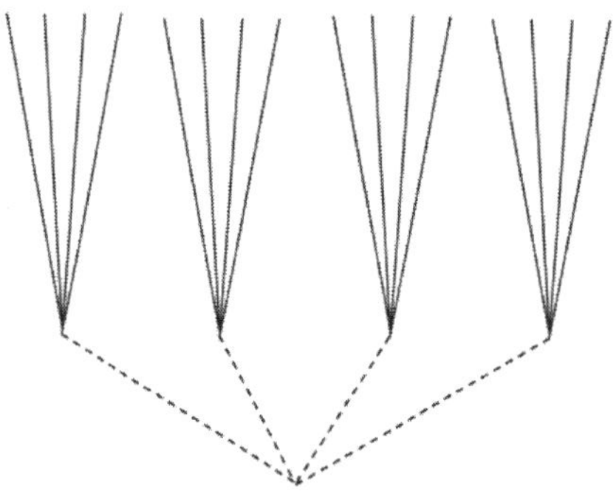

Figure 2. Joining 4 rooted Cayley tree of order 4 where r = 4 and k = 4.

Since the former C_r must always include x^0, its range should be from 1 to n. Total $D_k(n)$ is just the sum of all $C_k(r)C_k(n-r)$ using the addition and multiplication principles.

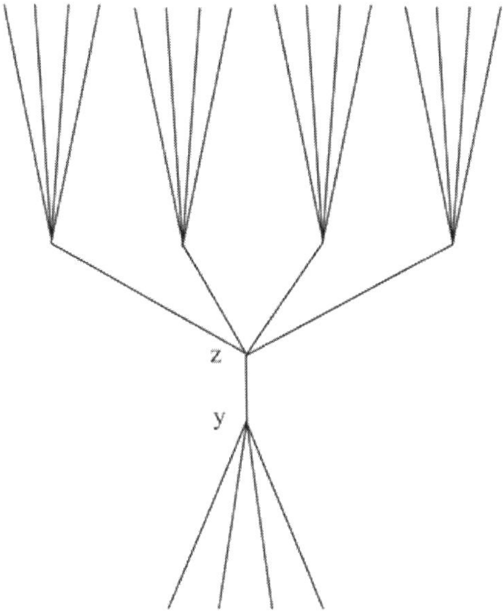

Figure 3. A homogenous graph where each vertex is connected to exactly 5 neighbours.

Using Equation (8), we rewrite the formula above as

$$D_k(n) = \sum_{r_1 + r_2 + \cdots + r_{k+1} = n-1}^{n} C_k(r_1) C_k(r_2) \cdots C_k(r_{k+1}) \quad \text{for } n > 0. \tag{12}$$

This formula can also be obtained using $k+1$ copies of k-ary tree together with $n-1$ vertices and one center.

We then find the binomial form of $D_k(n)$.

Corollary 4. For $k \geq 2$, $D_k(n)$ is expressed in binomial form as:

$$D_k(n) = R_k(n-1, k+1) = \frac{k+1}{(k-1)n+2}\binom{kn}{n-1} \quad \forall n > 0. \tag{13}$$

Thus, $D_2(n) = \frac{3}{n+2}\binom{2n}{n-1}$ and $D_3(n) = \frac{2}{n+2}\binom{3n}{n-1}$ as in [19] and [20], respectively. For k = 4, $D_4(n)$ coincides with one form of Raney numbers as mentioned above, i.e., $R_4(n,5)$. The numbers $D_k(n)$ generate a lot of new sequences. For example, the sequence of $R_5(n,6)$, i.e.,

$$1, 6, 45, 380, 3450, 32886, 324632, 3290040, 34034715, 357919100, \ldots$$

is not found in the OEIS database [6].

From theorem (2), one can get

$$D_k(n) = \sum_{r=0}^{n} C_k(r) C_k(n-r) - C_k(n).$$

This formula can be also obtained easily by a different way:

1) Count the number of trees by joining the 2 copies of $k-nary$ tree, with total number of vertices n.

2) Subtract those trees remunerate from y but doesn't contain z, that is, exactly the number $C_k(n)$.

Let the generating function of $D_k(n)$ be $w_k(x)$. Then we have the following result.

Corollary 5. For $k>1$ and $n\geq 0$, the generating function, $w_k(x)$ is

$$w_k(x)=u_k^2(x)-u_k(x) \tag{14}$$

where $u_k(x)$ is the generating function of $C_k(n)$.

Corollary 6. For $k>1$,

$$\sum_{r=1}^{n} C_k(r)C_k(n-r)=R_k(n-1,k+1) \tag{15}$$

Or

$$\sum_{r=1}^{n} R_k(r,1) R_k(n-r,1) = R_k(n-1,k+1). \tag{16}$$

Using the binomial inequality in [21] and the binomial forms of $C_k(n)$ and $D_k(n)$, the following in- equality can be easily proved.

Corollary 7. For $k > 2$,

$$C_k(n) \le D_k(n) < \frac{(k \cdot b_k)^n}{n^{3/2}} \tag{17}$$

where $b_k = \left(\frac{k}{k-1}\right)^{k-1}$ and $n > 0$.

For sufficiently large k, a simpler form is produced as well, i.e., $C_k(n) \le D_k(n) < \frac{(k \cdot e)^n}{n^{3/2}}$. These results are conjectured in a weaker form in [20], i.e., $C_k(n) \le D_k(n) < \frac{(k \cdot e)^n}{n}$.

4. CATALAN TRIANGLE

A Catalan triangle $B(n,r)$ is defined as follows [9] :

$$B(n,r)=\begin{cases}1 & \text{if } n=r=1;\\ B(n-1,r-1)+2B(n-1,r)+B(n-1,r+1) & \text{if } 1\le r\le n;\\ 0 & \text{otherwise.}\end{cases}$$

The Catalan triangle satisfies [9] :

$$B(n,r)=\sum_{\substack{i_1+\cdots+i_r=n\\ i_1,\cdots,i_r\ge 1}} C_2(i_1)C_2(i_2)\cdots C_2(i_r),\quad C_2(0)=1\quad \forall n>0 \tag{18}$$

Using a property of Catalan numbers, $C_2(i_r)=\sum_{j_1+j_2=i_r-1} C_2(j_1)C_2(j_2)$, where $j_1, j_2\ge 0$, we get another

form of $B(n,r)$, i.e.,

$$B(n,r)=\sum_{j_1+j_2+\cdots+j_r=n-r} C_2(j_1)C_2(j_2)\cdots C_2(j_{2r}), \tag{19}$$

where $j_1, j_2, \ldots, j_{2r} \in \mathbb{N} \cup \{0\}$. From Equation (1), we immediately recover the Catalan triangle from the Raney numbers, i.e., $B(n,r) = R_2(n-r, 2r)$:

$$B(n,r) = R_2(n-r, 2r) = \frac{r}{n}\binom{2n}{n-r}. \tag{20}$$

We now consider the following problem as in [22] : Find out the number of all different connected sub-trees of a homogenous binary tree with n number of vertices, containing the given r number of fixed vertices (where $n \geq 2r-2$). The condition, $n \geq 2r-2$, is simply the number of vertices that covers the minimal component containing all r vertices. The details of this problem and terminologies could be found in the original paper [22] . We denote the solution to this problem as F_n^r. In this paper, we show that a solution to the case when the minimal component is "full", is as below:

$$F_n^r = B(n-r+2, r) = \frac{r}{n-r+2}\binom{2(n-r+2)}{n+2},$$

where n is the number of given vertices and r is the number of fixed vertices in each of the connected sub- tree.

Now, we interpret and relate the problem above with the combinatorial interpretation of the Raney numbers through the following steps (see Figure 4):

1) Given n vertices;

2) Fill up all the interior points, i.e., $r-2$;

3) Fill up all the boundary points, i.e., r;

4) Then only $n-2r+2$ vertices are left;

5) Since each boundary point has 2 neighbours which is not an interior point, we have $2r$ boxes;

6) If $n-2r+2$ vertices are given, then there are $2r$ boxes of binary tree to be filled.

As a result, the solution is

$$R_2(n-2r+2,2r)=\frac{2r}{2(n-2r+2)+2r}\binom{2(n-2r+2)+2r}{n-2r+2}=\frac{r}{n-r+2}\binom{2(n-r+2)}{n+2}. \tag{21}$$

Furthermore, it is natural to define a generalized Catalan triangle, i.e., k-th Catalan triangle using Fuss- Catalan numbers instead of Catalan numbers as in Equation (19):

$$B_k(n,r) = \sum_{\substack{i_1 + \cdots + i_r = n \\ i_1, \cdots, i_r \geq 1}} C_k(i_1) C_k(i_2) \cdots C_k(i_r), \quad C_k(0) = 1 \quad \forall n > 0, \tag{22}$$

where $B_k(n,r) = 0$ if $n < r$.

From the property of Fuss-Catalan numbers, i.e., corollary (2)

$$C_k(i_r) = \sum_{j_1 + j_2 + \cdots + j_k = i_r - 1} C_k(j_1) C_k(j_2) \cdots C_k(j_k),$$

where $j_1, j_2, \cdots, j_k \geq 0$, we find another form of $B_k(n,r)$,

$$B_k(n,r) = \sum_{j_1 + j_2 + \cdots + j_r = n - r} C_k(j_1) C_k(j_2) \cdots C_k(j_{kr}), \tag{23}$$

where $j_1, j_2, \cdots, j_{kr} \in \mathbb{N} \cup \{0\}$. Again, from Equation (1), we immediately have

$$B_k(n,r)=R_k(n-r,kr)=\frac{r}{n}\binom{kn}{n-r}. \tag{24}$$

Lemma 2. Some properties of k-th Catalan triangles are as follows:

$$B_k(n,1)=C_k(n), \tag{25}$$

$$B_k(n,n)=1, \tag{26}$$

$$B_k(n,n-1)=(n-1)k. \tag{27}$$

Using the binomial form of $B_k(n,r)$, one can show that:

Lemma 3. For $k>1$ and $n\geq r\geq 2$

$$B_k(n-1,r-1)+2B_k(n-1,r)+B_k(n-1,r+1)=\frac{kr-k+2}{kn-k+2}\binom{kn-k+2}{n-r}. \tag{28}$$

If $k=2$, we recover

$$B_2(n-1,r-1)+2B_2(n-1,r)+B_2(n-1,r+1)=\frac{r}{n}\binom{2n}{n-r}=B_2(n,r).$$

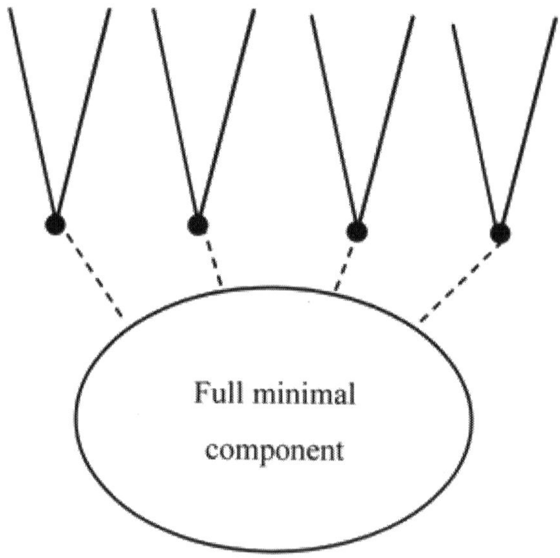

Figure 4. 4 boundary points (solid circles) connected to full minimal component.

Based on the initial result, lemma (3), we prove the following assertion by mathematical induction with respect to m.

Theorem 3. For fixed k, where $k \geq m > 1$, and $n \geq r > 1$,

$$\sum_{i=0}^{m}\binom{m}{i}B_k\left(n-1,r-1+i\right)=\frac{kr-k+m}{kn-k+m}\binom{kn-k+m}{n-r}, \tag{29}$$

where $B_k\left(n,r\right)=0$ if $n<r$.

Proof. Assertion is true for $m=2$. Assume that it is true for m, we consider the following summations:

$$\sum_{i=0}^{m}\binom{m}{i}B_k\left(n-1,r-1+i\right)+\sum_{i=0}^{m}\binom{m}{i}B_k\left(n-1,r+i\right)$$
$$=\frac{kr-k+m}{kn-k+m}\binom{kn-k+m}{n-r}+\frac{k\left(r+1\right)-k+m}{kn-k+m}\binom{kn-k+m}{n-r-1}.$$

$$B_k\left(n-1,r-1\right)+\sum_{i=1}^{m}\binom{m}{i}B_k\left(n-1,r-1+i\right)+\sum_{i=0}^{m-1}\binom{m}{i}B_k\left(n-1,r+i\right)+B_k\left(n-1,r+m\right)$$
$$=\frac{kr-k+\left(m+1\right)}{kn-k+\left(m+1\right)}\binom{kn-k+m+1}{n-r}.$$

$$B_k\left(n-1,r-1\right)+\sum_{i=1}^{m}\left(\binom{m}{i}+\binom{m}{i-1}\right)B_k\left(n-1,r-1+i\right)+B_k\left(n-1,r+m\right)$$
$$=\frac{kr-k+\left(m+1\right)}{kn-k+\left(m+1\right)}\binom{kn-k+m+1}{n-r}.$$

$$B_k(n-1,r-1)+\sum_{i=1}^{m}\binom{m+1}{i}B_k(n-1,r-1+i)+B_k(n-1,r+m)$$
$$=\frac{kr-k+(m+1)}{kn-k+(m+1)}\binom{kn-k+m+1}{n-r}.$$

$$\sum_{i=0}^{m+1}\binom{m+1}{i}B_k(n-1,r-1+i)=\frac{kr-k+(m+1)}{kn-k+(m+1)}\binom{kn-k+m+1}{n-r}.$$

Hence, the assertion is true for any $m>1$.

Corollary 8. For fixed k, where $k\geq m>1$, and $n\geq r>1$, we have

$$\sum_{i=0}^{m}\frac{r+i}{n}\binom{m}{i}\binom{kn}{n-r-i}=\frac{kr+m}{kn+m}\binom{kn+m}{n-r}. \tag{30}$$

For $m=k$, we have the following simple result:

Corollary 9. For fixed $k>1$, and $n\geq r>1$,

$$B_k(n,r)=\sum_{i=0}^{k}\binom{k}{i}B_k(n-1,r-1+i), \tag{31}$$

where $B_k(n,r)=0$ if $n<r$.

5. BINOMIAL TRANSFORMATION OF K-TH CATALAN TRIANGLE

For $r \geq 2$ and $n \geq r$, we define a new number $H_{k,m}^{(r)}(n)$ as:

$$H_{k,m}^{(r)}(n)=\sum_{\substack{j_1+\ldots+j_m+l_1+\ldots+l_r=n \\ j_1,\ldots,j_m \geq 0, k_1,\ldots,k_r \geq 1}} C_k(j_1)\ldots C_k(j_m)\cdot C_k(l_1)\ldots C_k(l_r).$$

(32)

If $j_1=j_2=\ldots=j_m=0$, all the Fuss-Catalan numbers should start at 1, then we recover the previously defined k-th Catalan triangle.

From the property of Fuss-Catalan numbers, $C_k(i_r)=\sum_{j_1+j_2+\ldots+j_k=i_r-1} C_k(j_1)C_k(j_2)\ldots C_k(j_k)$, where $j_1, j_2, \ldots \geq 0$, we found another form of

$$H_{k,m}^{(r)}(n)=\sum_{\substack{j_1+\ldots+j_m+l_1+\ldots+l_{kr}=n-r \\ j_1,\ldots,j_r,l_1,\ldots,l_{kr} \geq 0}} C_k(j_1)\ldots C_k(j_m)\cdot C_k(l_1)\ldots C_k(l_{kr}).$$

(33)

From Equation (1), we immediately have

$$H_{k,m}^{(r)}(n) = H_{k,m+kr}^{(0)}(n-r) = R_k(n-r, m+kr) = \frac{m+kr}{kn+m}\binom{kn+m}{n-r}$$

(34)

and for $m=0$, we recover the same formula for k-th Catalan triangle,

$$H_{k,m}^{(r)} = R_k(n-r, kr) = \frac{r}{n}\binom{kn}{n-r}.$$

From theorem (3), $H_{k,m}^{(r)}(n)$ is obtained as a result of binomial transformation of k-Catalan triangles.

Corollary 10. For fixed $k>1$, where $k \geq m > 1$, and $n \geq r > 1$,

$$\sum_{i=0}^{m}\binom{m}{i}B_k(n, r+i) = H_{k,m}^{(r)}(n),$$

(35)

where $B_k(n,r)=0$ if $n<r$.

6. CONCLUSION

In this paper, we have introduced the combinatorial interpretation of Raney numbers to solve various tree enumeration counting problems. The upper bound of any $k+1$ order tree enumeration is generally found to be $\frac{(k\cdot e)^n}{n^{3/2}}$. We have also shown how a new number $H_{k,m}^{(r)}(n)$ may be derived from the binomial transformation of th Catalan triangles.

ACKNOWLEDGEMENTS

This research is funded by the MOHE grant FRGS11-022-0170. The authors are grateful to anoymous referee's suggestion and improvement of the presentation of this paper.

REFERENCES

1. Penson, K.A. and Zyczkowski, K. (2011) Product of Ginibre Matrices: Fuss-Catalan and Raney Distributions. Physical Review E, 83, Article ID: 061118.

2. Mlotkowski, W., Penson, K.A. and Zyczkowski, K. (2013) Densities of the Raney Distributions. Documenta Mathematica, 18, 1573-1596.

3. Jeurissen, R.H. (2008) Raney and Catalan. Discrete Mathematics, 308, 6298-6307.

4. Dziemiaczuk, M. (2014) Enumerations of Plane Trees with Multiple Edges and Raney Lattice Paths. Discrete Mathematics, 337, 9-24.

5. Gould, H.W. (1972) Combinatorial Identities: A Standardized Set of Tables Listing 500 Binomial Coefficient Summations. Morgantown.

6. (2011) The On-Line Encyclopedia of Integer Sequences. Sequence A196678.http://oeis.org

7. Graham, R.L., Knuth, D.E. and Patashnik, O. (1994) Concrete Mathematics. Addison-Wesley, Boston.

8. Hilton, P. and Pedersen, J. (1991) Catalan Numbers, Their Generalization, and Their Uses. Mathematical Intelligencer, 13, 64-75.

9. Koshy, T. (2008) Catalan Numbers with Applications. Oxford University Press, USA.

10. Stanley, R.P. (2013) Catalan Addendum to Enumerative Combinatorics. Vol. 2.

11. Stanley, R.P. (1999) Enumerative Combinatorics. Vol. 2, Cambridge University Press, Cambridge.

12. Baxter, R.J. (1982) Exactly Solved Models in Statistical Mechanics. Academic Press, London.

13. Goltsev, A.V., Dorogovtsev, S.N. and Mendes, J.F.F. (2008) Critical Phenomena in Complex Networks. Reviews of Modern Physics, 80, 1275.

14. Tamm, U. (2001) Some Aspects of Hankel Matrices in Coding Theory and Combinatorics. The Electronic Journal of Combinatorics, 8, 1-31.

15. Lee, K., Jung, W.S., Park, J.S. and Choi, M.Y. (2000) Statistical Analysis of the Metropolitan Seoul Subway System: Network Structure and Passenger Flows. Physica A, 387, 6231-6234.

16. Aval, J.C. (2008) Multivariate Fuss-Catalan Numbers. Discrete Mathematics, 308, 4660-4669.

17. Shapiro, L.W. (1976) A Catalan Triangle. Discrete Mathematics, 14, 83-90.

18. Merlini, D., Sprugnoli, R. and Verri, M.C. (2006) Lagrange Inversion: When and How. Acta Applicandae Mathematica, 94, 233-249.

19. Pah, C.H. (2008) An Application of Catalan Number on Cayley Tree of Order 2: Single Polygon Counting. Bulletin of the Malaysian Mathematical Sciences Society, 31, 175-183.

20. Pah, C.H. (2010) Single Polygon Counting on Cayley Tree of Order 3. Journal of Statistical Physics, 140, 198-207.

21. Sasvari, Z. (1999) Inequalities for Binomial Coefficients. Journal of Mathematical Analysis and Applications, 236, 223-226.

22. Mukhomedov, F., Pah, C.H. and Saburov, M. (2010) Single Polygon Counting for m Fixed Nodes in Cayley Tree: Two Extremal Cases. Preprint. http://arxiv.org/abs/1004.2305

CHAPTER 5

Estimating the polygamma functions for negative integers

ABSTRACT

The polygamma functions $\psi(r)(x)$ are defined for all $x>0$ and $r \in N$. In this paper, the concepts of neutrix and neutrix limit are applied to generalize and redefine the polygamma functions $\psi(r)(x)$ for all $x \in R$ and $r \in N$. Also, further results are given.

KEYWORDS

gamma function; polygamma functions; neutrices; neutrix limits

1 INTRODUCTION

Neutrices are additive groups of negligible functions that do not contain any constants except zero. Their calculus was developed by van der Corput [1] and Hadamard in connection with asymptotic series and divergent integrals. Recently, the concepts of neutrix and neutrix limit have been used widely in many applications in mathematics, physics and statistics.

For example, Jack Ng and van Dam applied the neutrix calculus, in conjunction with the Hadamard integral developed by van der Corput, to the quantum field theories, in particular, to obtain finite results for the coefficients in the perturbation series. They also applied the neutrix calculus to quantum field theory and obtained finite renormalization in the loop calculations, see [2] and [3].

Further, many special functions have been generalized and redefined by using the neutrices. Fisher used the neutrices and the neutrix limit to define gamma, beta and incomplete gamma functions [4–8].

Ozcag *et al.* [9, 10] applied the neutrix limit to extend the definition of incomplete beta function and its partial derivatives for negative integers. Also, the digamma function was generalized for negative integers by Jolevska-Tuneska *et al.* [11]. Salem [12, 13] applied the neutrix limit to redefine the q-gamma and the incomplete q-gamma functions and their derivatives. In continuation, in this paper, we apply the concepts of neutrix and neutrix limit to generalize and redefine the polygamma functions.

A neutrix N is defined as a commutative additive group of functions $f(\xi)$ defined on a domain N' with values in an additive group N'', where further if for some f in N, $f(\xi)=\gamma$ for all $\xi \in N'$, then $\gamma=0$. The functions in N are called negligible functions.

Let N be a set contained in a topological space with a limit point a which does not belong to N. If $f(\xi)$ is a function defined on N' with values in N'' and it is possible to find a constant c such that $f(\xi)-c\in N$, then c is called the neutrix limit of f as ξ tends to a, and we write $N\text{-}\lim_{\xi\to a} f(\xi)=c$.

Note that if a neutrix limit c exists, then it is unique, since if $f(x)-c$ and $f(x)-c'$ are in N, then the constant function $c-c'$ is also in N and so $c=c'$.

Also note that if N is a neutrix containing the set of all functions which converge to zero in the normal sense as x tends to y, then

$$\lim_{x\to y} f(x) = c \quad \Rightarrow \quad N\text{-}\lim_{x\to y} f(x) = c.$$

In this paper, we let N be the neutrix having the domain $N'=\{\epsilon:0<\epsilon<\infty\}$ and the range N'' of real numbers, with negligible functions being finite linear sums of the functions

$$\epsilon^{\lambda} \ln^{r-1} \epsilon, \qquad \ln^{r} \epsilon \quad (\lambda < 0, r \in \mathbb{N})$$

and all functions $o(\epsilon)$ which converge to zero in the normal sense as ϵ tends to zero [1].

2 GAMMA AND DIGAMMA FUNCTIONS

The gamma function is defined as a locally summable function on the real line by [14]

$$\Gamma(x) = \int_0^{\infty} t^{x-1} e^{-x}\, dx, \quad x > 0. \tag{2.1}$$

In the classical sense, Γ(x) function is not defined for the negative integer thus still is an open problem to give satisfactory definition. However, by using the neutrix limit, it was shown in [6] that gamma function (2.1) is defined as follows:

$$\Gamma(x) = N\text{-}\lim_{\epsilon\to 0}\int_{\epsilon}^{\infty} t^{x-1}e^{-t}\,dt \tag{2.2}$$

for $x\neq 0,-1,-2,\ldots$, and this function is also defined by the neutrix limit

$$\begin{aligned}\Gamma(-m) &= N\text{-}\lim_{\epsilon\to 0}\int_{\epsilon}^{\infty} t^{-m-1}e^{-t}\,dt\\ &= \int_{1}^{\infty} t^{-m-1}e^{-t}\,dt + \int_{0}^{1} t^{-m-1}\left[e^{-t} - \sum_{i=0}^{m}\frac{(-1)^i}{i!}t^i\right]dt - \sum_{i=0}^{m-1}\frac{(-1)^i}{i!(m-i)}\end{aligned} \tag{2.3}$$

for $m\in$N. The existence of the r th derivative of the gamma function was also proven in [6] and defined by the equation

$$\begin{aligned}\Gamma^{(r)}(0) &= N\text{-}\lim_{\epsilon\to 0}\int_{\epsilon}^{\infty} t^{-1}\ln^r t e^{-t}\,dt\\ &= \int_{1}^{\infty} t^{-1}\ln^r t e^{-t}\,dt + \int_{0}^{1} t^{-1}\ln^r t\left[e^{-t}-1\right]dt,\end{aligned} \tag{2.4}$$

$$\begin{aligned}\Gamma^{(r)}(-m) &= N\text{-}\lim_{\epsilon\to 0}\int_{\epsilon}^{\infty} t^{-m-1}\ln^r t e^{-t}\,dt\\ &= \int_{1}^{\infty} t^{-m-1}\ln^r t e^{-t}\,dt + \int_{0}^{1} t^{-m-1}\ln^r t\left[e^{-t} - \sum_{i=0}^{m}\frac{(-1)^i}{i!}t^i\right]dt\\ &\quad - \sum_{i=0}^{m-1}\frac{(-1)^i}{i!}r!(m-i)^{-r-1}\end{aligned} \tag{2.5}$$

for $r \in N0$ and $m \in N$, see also [15]. Further, it was proven that

$$\Gamma(-r) = \frac{(-1)^r}{r!}\phi(r) - \frac{(-1)^r}{r!}\gamma \tag{2.6}$$

for r=1,2,... , where

$$\phi(r) = \sum_{i=1}^{r} \frac{1}{i},$$

thus we can extend the definition to the whole real line where

$$\Gamma(0) = \Gamma'(1) = -\gamma,$$

where γ denotes Euler's constant, see [16].

The logarithmic derivative of the gamma function is known as the *psi* or *digamma* function $\psi(x)$, that is, it is given by

$$\psi(x) = \frac{d}{dx}\left(\ln \Gamma(x)\right) = \frac{\Gamma'(x)}{\Gamma(x)}. \tag{2.7}$$

The digamma function is also defined as a locally summable function on the real line by

$$\psi(x) = -\gamma + \int_0^\infty \frac{e^{-t} - e^{-xt}}{1 - e^{-t}}\,dt, \quad x > 0, \tag{2.8}$$

and this can be read as

$$\psi(x) = -\gamma + \int_0^1 \frac{1 - t^{x-1}}{1 - t}\,dt, \quad x > 0, \tag{2.9}$$

where γ is also Euler's constant.

We note that the digamma function $\psi(x)$ is redefined by [11] to be

$$\psi(x) = -\gamma + H_n, \quad n \in \mathbb{N}_0, \tag{2.10}$$

and it has values for negative integers as

$$\psi(x) = -\gamma + H_n, \quad n \in \mathbb{N}_0, \tag{2.11}$$

where *Hn* denotes the Harmonic number defined as

$$H_n = \sum_{k=1}^{n} \frac{1}{k}, \quad n \in \mathbb{N} \text{ with } H_0 = 0.$$

It was also proven in [11] that

$$\psi'(-n) = \sum_{k=1}^{\infty} \frac{1}{(n-k)^2}, \quad n \in \mathbb{N}, \tag{2.12}$$

which is correct for all $n \neq k$; however, if $n=k$, then $\psi'(-n)$ is undefined, and we will give the exact formula for (2.12) in the following section.

3 POLYGAMMA FUNCTIONS

The polygamma functions $\psi(r)(x)$ are defined as the rth derivative of the digamma function, that is,

$$\psi^{(r)}(x) = -\int_0^1 \frac{t^{x-1} \ln^r t}{1-t}\, dt, \quad x > 0, r \in \mathbb{N}. \tag{3.1}$$

In the present section, we seek to redefine the polygamma functions $\psi^{(r)}(x)$ for all $x \in \mathrm{R}$ and $r \in \mathrm{N}$.

Lemma 3.1 *The neutrix limit, as ϵ tends to zero of the integral*

$$\int_\epsilon^1 t^{\alpha-1} \ln^r t\, dt, \tag{3.2}$$

exists for all values of $\alpha \in \mathrm{R}$ and $r \in \mathrm{N}$ and

$$N\text{-}\lim_{\epsilon\to 0}\int_\epsilon^1 t^{\alpha-1}\ln^r t\,dt = \begin{cases} \frac{(-1)^r r!}{\alpha^{r+1}}, & \alpha\neq 0,\\ 0, & \alpha = 0. \end{cases} \tag{3.3}$$

Proof: Let

$$I_r(a,b)=\int_a^b t^{\alpha-1}\ln^r t\,dt,\quad r\in\mathbb{N}.$$

It is easy to see by integration by parts that

$$I_r(a,b)=\frac{b^\alpha\ln^r b-a^\alpha\ln^r a}{\alpha}-\frac{r}{\alpha}I_{r-1}(a,b)\quad\text{and}\quad I_0(a,b)=\frac{b^\alpha-a^\alpha}{\alpha}.$$

The induction yields

$$I_r(a,b)=\frac{(-1)^r r!}{\alpha^{r+1}}\left(b^\alpha-a^\alpha\right)-\sum_{k=1}^{r}\frac{(-1)^r r!}{(r-k+1)!\alpha^k}\left(b^\alpha\ln^{r-k+1}b-a^\alpha\ln^{r-k+1}a\right).$$

Hence, we get

$$I_r(\epsilon,1)=\frac{(-1)^r r!}{\alpha^{r+1}}\left(1-\epsilon^\alpha\right)+\sum_{k=1}^{r}\frac{(-1)^r r!}{(r-k+1)!\alpha^k}\epsilon^\alpha\ln^{r-k+1}\epsilon.$$

Notice that the sum consists of (linear sum of $\epsilon^{\lambda}\ln^{k}\epsilon$) negligible functions which can be neglected when taking the neutrix limit as $\epsilon\to 0$. Taking the neutrix limit as $\epsilon\to 0$ yields the desired results.

Theorem 3.2 *Let $n,r\in\mathbb{N}$ and $x\in\mathbb{R}$. Then the neutrix limit, as ϵ tends to zero of the integral*

$$\int_{\epsilon}^{1} \frac{t^{x-1}\ln^{r} t}{1-t}\,dt, \tag{3.4}$$

exists for all values of $-n<x<-n+1$.

Proof Using the geometric sequence sum rule gives

$$\begin{aligned}\int_{\epsilon}^{1} \frac{t^{x-1}\ln^{r} t}{1-t}\,dt &= \int_{\epsilon}^{1} \frac{t^{x-1}\ln^{r} t(1-t^{n})}{1-t}\,dt + \int_{\epsilon}^{1} \frac{t^{x+n-1}\ln^{r} t}{1-t}\,dt \\ &= \sum_{k=0}^{n-1}\int_{\epsilon}^{1} t^{x+k-1}\ln^{r} t\,dt + \int_{\epsilon}^{1} \frac{t^{x+n-1}\ln^{r} t}{1-t}\,dt.\end{aligned}$$

Taking the neutrix limit as $\epsilon\to 0$ yields

$$N\text{-}\lim_{\epsilon\to 0}\int_{\epsilon}^{1} \frac{t^{x-1}\ln^{r} t}{1-t}\,dt = \sum_{k=0}^{n-1}\left(N\text{-}\lim_{\epsilon\to 0}\int_{\epsilon}^{1} t^{x+k-1}\ln^{r} t\,dt\right) + \int_{0}^{1} \frac{t^{x+n-1}\ln^{r} t}{1-t}\,dt.$$

Lemma 3.1 indicates that the neutrix limit of the integrals

$$N\text{-}\lim_{\epsilon\to 0}\int_{\epsilon}^{1} t^{x+k-1}\ln^{r} t\,dt$$

exist for all $-n<x<-n+1$ and thus integral (3.4) also exists. □

Theorem 3.3 *The neutrix limit, as* ϵ *tends to zero of the integral*

$$\int_{\epsilon}^{1}\frac{t^{-1}\ln^{r} t}{1-t}\,dt, \tag{3.5}$$

exists for all $r\in\mathbb{N}$ *and*

$$N\text{-}\lim_{\epsilon\to 0}\int_{\epsilon}^{1}\frac{t^{-1}\ln^{r} t}{1-t}\,dt=-\psi^{(r)}(1). \tag{3.6}$$

Proof We have

$$\int_{\epsilon}^{1}\frac{t^{-1}\ln^{r} t}{1-t}\,dt=\int_{\epsilon}^{1} t^{-1}\ln^{r} t\,dt+\int_{\epsilon}^{1}\frac{\ln^{r} t}{1-t}\,dt.$$

In view of Lemma 3.1, the neutrix limit of the first integral exists and equals zero, and so we get

$$N\text{-}\lim_{\epsilon\to 0}\int_{\epsilon}^{1}\frac{t^{-1}\ln^{r} t}{1-t}\,dt=\int_{0}^{1}\frac{\ln^{r} t}{1-t}\,dt=-\psi^{(r)}(1).$$

Theorem 3.4 *The neutrix limit, as ϵ tends to zero of the integral*

$$\int_{\epsilon}^{1}\frac{t^{-n-1}\ln^{r} t}{1-t}\,dt, \tag{3.7}$$

exists for all n,r∈N and

$$N\text{-}\lim_{\epsilon\to 0}\int_{\epsilon}^{1}\frac{t^{-n-1}\ln^{r} t}{1-t}\,dt=-\psi^{(r)}(1)-\sum_{k=1}^{n}\frac{r!}{k^{r+1}}. \tag{3.8}$$

Proof Using the geometric sequence sum rule gives

$$\begin{aligned}\int_{\epsilon}^{1}\frac{t^{-n-1}\ln^{r} t}{1-t}\,dt &=\int_{\epsilon}^{1}\frac{t^{-n-1}(1-t^{n+1})\ln^{r} t}{1-t}\,dt+\int_{\epsilon}^{1}\frac{\ln^{r} t}{1-t}\,dt\\ &=\sum_{k=0}^{n}\int_{\epsilon}^{1}t^{k-n-1}\ln^{r} t\,dt+\int_{\epsilon}^{1}\frac{\ln^{r} t}{1-t}\,dt\\ &=\sum_{k=0}^{n-1}\int_{\epsilon}^{1}t^{k-n-1}\ln^{r} t\,dt+\int_{\epsilon}^{1}t^{-1}\ln^{r} t\,dt+\int_{\epsilon}^{1}\frac{\ln^{r} t}{1-t}\,dt\\ &=\sum_{k=1}^{n}\int_{\epsilon}^{1}t^{-k-1}\ln^{r} t\,dt+\int_{\epsilon}^{1}t^{-1}\ln^{r} t\,dt+\int_{\epsilon}^{1}\frac{\ln^{r} t}{1-t}\,dt.\end{aligned}$$

Taking the neutrix limit as $\epsilon \to 0$ yields

$$N\text{-}\lim_{\epsilon\to 0}\int_{\epsilon}^{1}\frac{t^{-n-1}\ln^{r} t}{1-t}\,dt = \sum_{k=1}^{n}\left(N\text{-}\lim_{\epsilon\to 0}\int_{\epsilon}^{1} t^{-k-1}\ln^{r} t\,dt\right)$$
$$+ N\text{-}\lim_{\epsilon\to 0}\int_{\epsilon}^{1} t^{-1}\ln^{r} t\,dt + \int_{0}^{1}\frac{\ln^{r} t}{1-t}\,dt.$$

The results obtained in Lemma 3.1 and definition (3.1) complete the proof.

The above theorems lead us to introducing the following.

Definition 3.5 The polygamma function $\psi^{(r)}(x)$ can be redefined by

$$\psi^{(r)}(x) = -\left(N\text{-}\lim_{\epsilon\to 0}\int_{\epsilon}^{1}\frac{t^{x-1}\ln^{r} t}{1-t}\,dt\right) \tag{3.9}$$

for all $x \in \mathrm{R}$ and $r \in \mathrm{N}$.

Corollary 3.6 *The polygamma function $\psi(r)(x)$ has value at $x=0$ as*

$$\psi^{(r)}(0) = \psi^{(r)}(1) \tag{3.10}$$

and for negative integers x=−n as

$$\psi^{(r)}(-n) = \psi^{(r)}(1) + \sum_{k=1}^{n} \frac{r!}{k^{r+1}}, \quad n \in \mathbb{N}. \tag{3.11}$$

In particular, when r=1, we have

$$\psi'(0) = \psi'(1) = \frac{\pi^2}{6} \simeq 1.64493 \tag{3.12}$$

And

$$\psi'(-n) = \psi'(1) + \sum_{k=1}^{n} \frac{1}{k^2} = \frac{\pi^2}{6} + \sum_{k=1}^{n} \frac{1}{k^2} \simeq 1.64493 + \sum_{k=1}^{n} \frac{1}{k^2}, \quad n \in \mathbb{N}. \tag{3.13}$$

Corollary 3.7 *If we let n→∞ to* (3.11), *then we get*

$$\psi^{(r)}(-\infty) = \psi^{(r)}(1) + r!\zeta(r+1), \quad r \in \mathbb{N}, \tag{3.14}$$

where ζ(s) is the Riemann zeta function defined as [14]

$$\zeta(s) = \sum_{k=1}^{\infty} \frac{1}{k^s}, \quad \Re(s) > 1.$$

Since $\psi^{(r)}(1) = (-1)^{r+1} r!\zeta(r+1)$,*, then we get*

$$\psi^{(r)}(-\infty) = r!\zeta(r+1)\left(1+(-1)^{r+1}\right) = \begin{cases} 2r!\zeta(r+1) & \textit{if n is odd,} \\ 0 & \textit{if n is even,} \end{cases} \tag{3.15}$$

and if r=1, we get

$$\psi'(-\infty) = 2\zeta(2) = \frac{\pi^2}{3} \simeq 3.28987. \tag{3.16}$$

Remark 3.8 Formula (3.13) is the correct formula of (2.12).

ACKNOWLEDGEMENTS

The authors would like to thank the referee(s) for valuable remarks and suggestions on the previous version of the manuscript.

COMPETING INTERESTS

The authors declare that they do not have competing interests in publishing this article.

AUTHORS' CONTRIBUTIONS

Both authors contributed equally. All authors read and approved the final manuscript.

REFERENCES

1. van der Corput JG: Introduction to the neutrix calculus. *J. Anal. Math.* 1960, 7: 291–398.

2. Ng YJ, van Dam H: Neutrix calculus and finite quantum field theory. *J. Phys. A* 2005, 38(18):L317-L323.

3. Ng YJ, van Dam H: An application of neutrix calculus to quantum theory. *Int. J. Mod. Phys. A* 2006, 21(2):297–312.

4. Fisher B:On defining the incomplete gamma function $\gamma(-m,x)$. *Integral Transforms Spec. Funct.* 2004, 15(6):467–476. 10.1080/10652460412331270698

5. Fisher B, Jolevska-Tuneska B, Kılıçman A: On defining the incomplete gamma function. *Integral Transforms Spec. Funct.* 2003, 14(4):293–299.

6. Fisher B, Kılıçman A, Nicholas D: On the beta function and the neutrix product of distributions. *Integral Transforms Spec. Funct.* 1998, 7(1–2):35–42.

7. Fisher B, Kuribayashi Y: Neutrices and the beta function. *Rostock. Math. Kolloqu.* 1987, 32: 56–66.

8. Fisher B, Kuribayashi Y: Neutrices and the gamma function. *J. Fac. Ed. Tottori Univ. Math. Sci.* 1987, 36(1–2):1–7.

9. Ozcag E, Ege I, Gurcay H: On partial derivative of the incomplete beta function. *Appl. Math. Lett.* 2008, 21: 675–681. 10.1016/j.aml.2007.07.020

10. Ozcag E, Ege I, Gurcay H: An extension of the incomplete beta function for negative integers. *J. Math. Anal. Appl.* 2008, 338: 984–992. 10.1016/j.jmaa.2007.05.075

11. Jolevska-Tuneska B, Jolevski I: Some results on the digamma function. *Appl. Math. Inform. Sci.* 2013, 7(1):167–170.

12. Salem A: The neutrix limit of the q -gamma function and its derivatives. *Appl. Math. Lett.* 2010, 23: 1262–1268.

13. Salem A: Existence of the neutrix limit of the q -analogue of the incomplete gamma function and its derivatives. *Appl. Math. Lett.* 2012, 25: 363–368.

14. Abramowitz M, Stegun CA National Bureau of Standards Applied Mathematics Series 55. In *Handbook of Mathematical Functions with Formulas, Graphs, Mathematical Tables*. National Bureau of Standards, Washington; 1964. 7th printing

15. Ozcag E, Ege I, Gurcay H, Jolevska-Tuneska B: Some remarks on the incomplete gamma function. In *Mathematical Methods in Engineering*. Springer, Dordrecht; 2007:97–108.

16. Fisher B, Kılıçman A: Some results on the gamma function for negative integers. *Appl. Math. Inform. Sci.* 2012, 6(2):173–176.

CHAPTER 6

A family of generalized Stirling numbers of the first kind

Beih S. El-Desouky[1], Nabela A. El-Bedwehy[2], Abdelfattah Mustafa[1], Fatma M. Abdel Menem[2]

[1]Mathematics Department, Faculty of Science, Mansoura University, Mansoura, Egypt

[2]Mathematics Department, Faculty of Science, Damietta University, Damietta, Egypt

ABSTRACT

A modified approach via differential operator is given to derive a new family of generalized Stirling numbers of the first kind. This approach gives us an extension of the techniques given by El-Desouky [1] and Gould [2] . Some new combinatorial identities and many relations between different types of Stirling numbers are found. Furthermore, some interesting special cases of the generalized Stirling numbers of the first kind are deduced. Also, a connection between these numbers and the generalized harmonic numbers is derived. Finally, some applications in coherent states and matrix representation of some results obtained are given.

KEYWORDS

Stirling numbers, Comtet numbers, Creation, Annihilation, Differential operator, Maple program

1. INTRODUCTION

Gould [2] proved that

$$\left(e^{x}D\right)^{n}=e^{nx}\sum_{k=1}^{n}(-1)^{n-k}s(n,k)D^{k}=e^{nx}\sum_{k=1}^{n}s_{1}(n,k)D^{k},D=D_{x}=\frac{d}{dx},\tag{1}$$

where $s(n,k)$ and $s_1(n,k)$ are the usual Stirling numbers and the singles Stirling numbers of the first kind, respectively, defined by

$$(x)_{n}=\sum_{k=0}^{n}s(n,k)x^{k},s(n,0)=\delta_{n,0}\text{ and }s(n,k)=0\text{ for }k>n.\tag{2}$$

$$\langle x\rangle_{n}=\sum_{k=0}^{n}s_{1}(n,k)x^{k},s_{1}(n,0)=\delta_{n,0}\text{ and }s_{1}(n,k)=0\text{ for }k>n,\tag{3}$$

where $(x)_n = x(x-1)\cdots(x-n+1)$ and.

These numbers satisfy the recurrence relations

$$s(n+1,k)=s(n,k-1)-ns(n,k),\tag{4}$$

$$s_{1}(n+1,k)=s_{1}(n,k-1)+ns_{1}(n,k).\tag{5}$$

EL-Desouky [1] defined the generalized Stirling numbers of the first kind $s(k;\overline{\boldsymbol{r}},\overline{\boldsymbol{s}})$, called $(\overline{\boldsymbol{r}},\overline{\boldsymbol{s}})$-Stirling numbers of the first kind by

$$e^{r_n x} D^{s_n} \cdots e^{r_2 x} D^{s_2} e^{r_1 x} D^{s_1} = e^{\left(\sum_{i=1}^{n} r_i\right)x} \sum_{k=s_1}^{s_1+s_2+\cdots+s_n} s(k;\overline{\boldsymbol{r}},\overline{\boldsymbol{s}}) D^k, \tag{6}$$

$s(k;\overline{\boldsymbol{r}},\overline{\boldsymbol{s}})=0$ for $k<s_1$ or $k>\sum_{j=1}^{n} s_j$ and $s(0;\overline{\boldsymbol{r}},\boldsymbol{s})=1$, where $\overline{\boldsymbol{r}}:=(r_1,r_2,\cdots,r_n)$ is a sequence of real numbers and $\overline{\boldsymbol{s}}:=(s_1,s_2,\cdots,s_n)$ is a sequence of nonnegative integers.

Equation (6) is equivalent to

$$e^{r_n a^+} a^{s_n} \cdots e^{r_2 a^+} a^{s_2} e^{r_1 a^+} a^{s_1} = e^{\left(\sum_{i=1}^{n} r_i\right)a^+} \sum_{k=s_1}^{s_1+s_2+\cdots+s_n} s(k;\overline{\boldsymbol{r}},\overline{\boldsymbol{s}}) a^k, \tag{7}$$

where a^+ and a are boson creation and annihilation operators, respectively, and satisfy the commutation relation $\left[a^+,a\right]=1.$

The numbers $s(k;\overline{\boldsymbol{r}},\overline{\boldsymbol{s}})$ satisfy the recurrence relation

$$s(k;\overline{\boldsymbol{r}}\oplus r_{n+1},\overline{\boldsymbol{s}}\oplus s_{n+1}) = \sum_{i=0}^{s_{n+1}} \binom{s_{n+1}}{i} \left(\sum_{j=1}^{n} r_j\right)^{s_{n+1}-i} s(k-i;\overline{\boldsymbol{r}},\overline{\boldsymbol{s}}), \tag{8}$$

with the notations $\overline{\boldsymbol{r}}\oplus r_{n+1}=(r_1,r_2,\cdots,r_{n+1})$ and $\overline{\boldsymbol{s}}\oplus s_{n+1}=(s_1,s_2,\cdots,s_{n+1})$.

The numbers $s(k;\bar{r},\bar{s})$ have the explicit formula

$$s(k;\bar{r},\bar{s})=\sum_{\sigma_{n-1}=\beta_n-k,j\geq 0}\prod_{l=1}^{n-1}\binom{s_{l+1}}{i_l}\left(\sum_{j=1}^{l}r_j\right)^{i_l}, \tag{9}$$

where $\sigma_n=\sum_{j=0}^{n}i_j$, with $i_0=0$ and $\beta_n=\sum_{j=1}^{n}s_j$.

Moreover El-Desouky [1] derived many special cases and some applications. For the proofs and more details, see [1] .

The generalized falling factorial of x associated with the sequence $\bar{\alpha}:=(\alpha_0,\alpha_1,\cdots,\alpha_{n-1})$ of order n, where $\alpha_0,\alpha_1,\cdots,\alpha_{n-1}$ are real numbers, is defined by $(x,\bar{\alpha})_n=\prod_{i=0}^{n-1}(x-\alpha_i),(x,\bar{\alpha})_0=1.$

Comtet [3] [4] and [5] defined $s_{\bar{\alpha}}(n,k)$ the generalized Stirling numbers of the first kind, which are called Comtet numbers, by

$$(x,\bar{\alpha})_n=\sum_{k=0}^{n}s_{\bar{\alpha}}(n,k)x^k. \tag{10}$$

These numbers satisfy the recurrence relation

$$s_{\bar{\alpha}}(n,k)=s_{\bar{\alpha}}(n-1,k-1)-\alpha_{n-1}s_{\bar{\alpha}}(n-1,k). \tag{11}$$

El-Desouky and Cakic [6] defined $s_{\bar{\alpha}}(n,k;\bar{s})$, the generalized Comtet numbers by

$$\prod_{j=0}^{n-1}\left(x^{\alpha_j}\delta x^{-\alpha_j}\right)^{s_j}=\prod_{j=0}^{n-1}\left(\delta-\alpha_j\right)^{s_j}=\sum_{k=0}^{|s|}s_{\bar{\alpha}}\left(n,k;\tilde{s}\right)\delta^k, \tag{12}$$

where $|s|=\sum_{j=0}^{n-1}s_j, s_{\bar{\alpha}}\left(n,k;s\right)=0$ for $k>|s|, s_{\bar{\alpha}}\left(n,0;\tilde{s}\right)=(-1)^{|s|}\prod_{j=0}^{n-1}\left(\alpha_j\right)^{s_j}$ and $\tilde{s}=\left(s_0,s_1,\cdots,s_n\right)$.

For more details on generalized Stirling numbers via differential operators, see [7] - [10] and [11] .

The paper is organized as follows:

In Section 2, using the differential operator $\left(e^{r_n x}D^{s_n}e^{r_n x}\right)\cdots\left(e^{r_2 x}D^{s_2}e^{r_2 x}\right)\left(e^{r_1 x}D^{s_1}e^{r_1 x}\right)$ we define a new family of generalized Stirling numbers of the first kind, denoted by $s\left(k;\bar{r},\bar{s}\right)$. A recurrence relation and an explicit formula of these numbers are derived. In Section 3, some interesting special cases are discussed. Moreover some new combinatorial identities and a connection between $s\left(n,k;r,s\right)$ and the generalized harmonic numbers $O_n^{(i)}$ are given. In Section 4, some applications in coherent states and matrix representation of some results obtained are given. Section 5 is devoted to the conclusion, which handles the main results derived throughout this work. Finally, a computer program is written using Maple and executed for calculating the generalized Stirling numbers of the first kind and some special cases, see Appendix.

2. MAIN RESULTS

Let $\overline{\boldsymbol{r}} := (r_1, r_2, \cdots, r_n)$ be a sequence of real numbers and $\overline{\boldsymbol{s}} := (s_1, s_2, \cdots, s_n)$ be a sequence of nonnegative integers.

Definition 2.1

The generalized Stirlng numbers $\boldsymbol{s}(k; \overline{\boldsymbol{r}}, \overline{\boldsymbol{s}})$ are defined by

$$\left(\mathrm{e}^{r_n x} D^{s_n} \mathrm{e}^{r_n x}\right) \cdots \left(\mathrm{e}^{r_2 x} D^{s_2} \mathrm{e}^{r_2 x}\right)\left(\mathrm{e}^{r_1 x} D^{s_1} \mathrm{e}^{r_1 x}\right) = \mathrm{e}^{\left(2\sum_{l=1}^{n} r_l\right)} \sum_{k=0}^{\beta_n} \boldsymbol{s}(k; \overline{\boldsymbol{r}}, \overline{\boldsymbol{s}}) D^k, \tag{13}$$

where $\boldsymbol{s}(k; \overline{\boldsymbol{r}}, \overline{\boldsymbol{s}}) = 0$ $\beta_n = \sum_{j=1}^{n} s_j,$ for $k > \beta_n$ and $\boldsymbol{s}(0; \overline{\boldsymbol{r}}, \overline{\boldsymbol{s}}) = 1$.

Equation (13) is equivalent to

$$\left(\mathrm{e}^{r_n a^+} a^{s_n} \mathrm{e}^{r_n a^+}\right) \cdots \left(\mathrm{e}^{r_2 a^+} a^{s_2} \mathrm{e}^{r_2 a^+}\right)\left(\mathrm{e}^{r_1 a^+} a^{s_1} \mathrm{e}^{r_1 a^+}\right) = \mathrm{e}^{\left(2\sum_{l=1}^{n} r_l\right)a^+} \sum_{k=0}^{\beta} \boldsymbol{s}(k; \overline{\boldsymbol{r}}, \overline{\boldsymbol{s}}) a^k. \tag{14}$$

Theorem 2.1

The numbers $\boldsymbol{s}(k; \overline{\boldsymbol{r}}, \overline{\boldsymbol{s}})$ satisfy the recurrence relation

$$\boldsymbol{s}(k; \overline{\boldsymbol{r}} \oplus r_{n+1}, \overline{\boldsymbol{s}} \oplus s_{n+1}) = \sum_{i=0}^{s_{n+1}} \binom{s_{n+1}}{i} \left(2\sum_{j=1}^{n} r_j + r_{n+1}\right)^{s_{n+1}-i} \boldsymbol{s}(k-i; \overline{\boldsymbol{r}}, \overline{\boldsymbol{s}}), \tag{15}$$

with the notations

$\overline{\boldsymbol{r}} \oplus r_{n+1} := (r_1, r_2, \cdots, r_{n+1})$ and $\overline{\boldsymbol{s}} \oplus s_{n+1} := (s_1, s_2, \cdots, s_{n+1})$.

Proof

$$
\begin{aligned}
e^{\left(2\sum_{j=1}^{n+1}r_j\right)x}\sum_{k=0}^{\beta_{n+1}}\boldsymbol{s}(k;\overline{\boldsymbol{r}},\overline{\boldsymbol{s}})D^k &= \left(e^{r_{n+1}x}D^{s_{n+1}}e^{r_{n+1}x}\right)\left(e^{\left(2\sum_{j=1}^{n}r_j\right)x}\sum_{k=0}^{\beta_n}\boldsymbol{s}(k;\overline{\boldsymbol{r}},\overline{\boldsymbol{s}})D^k\right)\\
&= e^{r_{n+1}x}D^{s_{n+1}}\left(e^{\left(2\sum_{j=1}^{n}r_j+r_{n+1}\right)x}\sum_{k=0}^{\beta_n}\boldsymbol{s}(k;\overline{\boldsymbol{r}},\overline{\boldsymbol{s}})D^k\right)\\
&= e^{\left(2\sum_{j=1}^{n}r_j\right)x}\sum_{m=0}^{\beta_n}\left(D+\left(2\sum_{j=1}^{n}r_j+r_{n+1}\right)\right)^{s_{n+1}}\boldsymbol{s}(m;\overline{\boldsymbol{r}},\overline{\boldsymbol{s}})D^m\\
&= e^{\left(2\sum_{j=1}^{n}r_j\right)x}\sum_{m=0}^{\beta_n}\sum_{i=0}^{s_{n+1}}\binom{s_{n+1}}{i}\left(2\sum_{j=1}^{n}r_j+r_{n+1}\right)^{s_{n+1}-i}\boldsymbol{s}(m;\overline{\boldsymbol{r}},\overline{\boldsymbol{s}})D^{m+i}\\
&= e^{\left(2\sum_{j=1}^{n}r_j\right)x}\sum_{k=0}^{\beta_{n+1}}\sum_{i=0}^{s_{n+1}}\binom{s_{n+1}}{i}\left(2\sum_{j=1}^{n}r_j+r_{n+1}\right)^{s_{n+1}-i}\boldsymbol{s}(k-i;\overline{\boldsymbol{r}},\overline{\boldsymbol{s}})D^k.
\end{aligned}
$$

Equating the coefficients of D^k on both sides yields (15).

Theorem 2.2

The numbers $\boldsymbol{s}(k;\overline{\boldsymbol{r}},\overline{\boldsymbol{s}})$ have the explicit formula

$$
\boldsymbol{s}(k;\overline{\boldsymbol{r}},\overline{\boldsymbol{s}})=\sum_{\sigma_n=\beta_n-k,\,j\ge 0}\prod_{l=1}^{n}\binom{s_i}{i}\left(2\left(\sum_{j=0}^{l-1}r_j\right)+r_l\right)^{i_l},\ \text{where}\ \sigma_n=\sum_{j=1}^{n}i_j,\ r_0=0. \tag{16}
$$

Proof

$$
e^{r_1x}D^{s_1}e^{r_1x}=e^{r_1x}e^{r_1x}\left(D+r_1I\right)^{s_1}=e^{2r_1x}\sum_{i_1=0}^{s_1}\binom{s_1}{i_1}(r_1)^{i_1}D^{s_1-i_1},
$$

$$
\begin{aligned}
\left(e^{r_2x}D^{s_2}e^{r_2x}\right)\left(e^{r_1x}D^{s_1}e^{r_1x}\right) &= e^{r_2x}D^{s_2}e^{r_2x}\left(e^{2r_1x}\sum_{i_1=0}^{s_1}\binom{s_1}{i_1}(r_1)^{i_1}D^{s_1-i_1}\right)=e^{r_2x}D^{s_2}e^{(2r_1+r_2)x}\sum_{i_1=0}^{s_1}\binom{s_1}{i_1}(r_1)^{i_1}D^{s_1-i_1}\\
&= e^{r_2x}e^{(2r_1+r_2)x}\sum_{i_1=0}^{s_1}\binom{s_1}{i_1}(r_1)^{i_1}\left(D+(2r_1+r_2)I\right)^{s_2}D^{s_1-i_1}\\
&= e^{2(r_1+r_2)x}\sum_{i_1=0}^{s_1}\sum_{i_2=0}^{s_2}\binom{s_1}{i_1}\binom{s_2}{i_2}(r_1)^{i_1}(2r_1+r_2)^{i_2}D^{s_1+s_2-i_1-i_2},
\end{aligned}
$$

thus, by iteration, we get

$$(e^{r_n x}D^{s_n}e^{r_n x})\cdots(e^{r_2 x}D^{s_2}e^{r_2 x})(e^{r_1 x}D^{s_1}e^{r_1 x})$$
$$=e^{2\left(\sum_{l=1}^{n} r_l\right)x}\sum_{i_1=0}^{s_1}\sum_{i_2=0}^{s_2}\cdots\sum_{i_n=0}^{s_n}\prod_{j=1}^{n}\binom{s_j}{i_j}\left(2\left(\sum_{l=1}^{j-1} r_l\right)+r_j\right)^{i_j} D^{\sum_{j=1}^{n} s_j-\sum_{j=1}^{n} i_j}, r_0=0. \tag{17}$$

Setting $\sum_{j=1}^{n} s_j-\sum_{j=1}^{n} i_j=\beta_n-\sigma_n=k,$ we obtain

$$\left(e^{r_n x}D^{s_n}e^{r_n x}\right)\cdots\left(e^{r_2 x}D^{s_2}e^{r_2 x}\right)\left(e^{r_1 x}D^{s_1}e^{r_1 x}\right)=e^{\left(2\sum_{l=1}^{n} r_l\right)x}\sum_{k=0}^{\beta_n}\sum_{\sigma_n=\beta_n-k, j\ge 0}\prod_{j=1}^{n}\binom{s_j}{i_j}\left(2\left(\sum_{l=1}^{j-1} r_l\right)+r_j\right)^{i_j} D^k. \tag{18}$$

Comparing (13) and (18) yields (16).

3. SPECIAL CASES

Setting $r_i=r$ and $s_i=s, i=1,\cdots,n$ in (13), we have the following definition.

Definition 3.1

For any real number r and nonnegative integer s, let the numbers $\boldsymbol{s}(n,k;r,s)$, be defined by

$$\left(e^{rx}D^s e^{rx}\right)^n=e^{(2nr)x}\sum_{k=0}^{ns}\boldsymbol{s}(n,k;r,s)D^k, \tag{19}$$

where $\boldsymbol{s}(0,0;r,s)=1$ and $\boldsymbol{s}(n,k;r,s)=0$ for $k>ns$.

Equation (19) is equivalent to

$$\left(e^{ra^+} a^s e^{ra^+}\right)^n = e^{(2nr)a^+} \sum_{k=0}^{ns} \boldsymbol{s}(n,k;r,s) a^k. \tag{20}$$

Corollary 3.1

The numbers $\boldsymbol{s}(n,k;r,s)$ satisfy the recurrence relation

$$\boldsymbol{s}(n+1,k;r,s) = \sum_{i=0}^{s} \binom{s}{i} \left((2n+1)r\right)^{s-i} \boldsymbol{s}(n,k-i;r,s). \tag{21}$$

Proof

The proof follows directly from equation (15) by setting $r_i = r$ and $s_i = s, i = 1,2,\cdots,n.$

Corollary 3.2

The numbers $\boldsymbol{s}(n,k;r,s)$ have the explicit formula

$$\boldsymbol{s}(n,k;r,s) = r^{ns-k} \sum_{\sigma_n = ns-k, i_j \geq 0} \left(\prod_{j=1}^{n} \binom{s}{i_j} (2j-1)^{i_j} \right). \tag{22}$$

Proof

By substituting $r_i = r$ and $s_i = s, i = 1,2,\cdots,n$ in Equation (17), yields

$$\left(e^{rx} D^s e^{rx}\right)^n = e^{(2nr)x} \sum_{0 \leq i_1,\cdots,i_n \leq s} \left(\prod_{j=1}^{n} \binom{s}{i_j} (2j-1)^{i_j} r^{\sigma_n} \right) D^{ns-\sigma_n},$$

then setting $ns-\sigma_n=k$ we have

$$\left(e^{rx}D^s e^{rx}\right)^n = e^{(2nr)x}\sum_{k=0}^{ns} r^{ns-k}\sum_{\sigma_n=ns-k,i_j\geq 0}\left(\prod_{j=1}^{n}\binom{s}{i_j}(2j-1)^{i_j}\right)D^k, \tag{23}$$

hence comparing equations (19) and (23) we obtain equation (22).

Furthermore we handle the following special cases.

i) If $r=1$, then we have

Definition 3.2

$$\left(e^x D^s e^x\right)^n = e^{(2n)x}\sum_{k=0}^{ns} \boldsymbol{s}(n,k;1,s)D^k, \tag{24}$$

where $\boldsymbol{s}(0,0;1,s)=1$ and $\boldsymbol{s}(n,k;1,s)=0$ for $k>ns$.

Corollary 3.3

The numbers $\boldsymbol{s}(n,k;1,s)$ satisfy the recurrence relation

$$\boldsymbol{s}(n+1,k;1,s)=\sum_{i=0}^{s}\binom{s}{i}(2n+1)^{s-i}\,\boldsymbol{s}(n,k-i;1,s). \tag{25}$$

Proof:

The proof follows directly from Equation (21) by setting $r=1$.

Corollary 3.4

The numbers $\boldsymbol{s}(n,k;1,s)$ have the explicit formula

$$\boldsymbol{s}(n,k;1,s)=\sum_{\sigma_n=ns-k,i_j\geq 0}\left[\prod_{j=1}^{n}\binom{s}{i_j}(2j-1)^{i_j}\right]. \tag{26}$$

Proof

The proof follows directly from Equation (22) by setting $r=1$.

ii) If $s=1$, then we have

Definition 3.3

The numbers $\boldsymbol{s}(n,k;r,1)$ are defined by

$$\left(e^{rx}De^{rx}\right)^n=e^{(2nr)x}\sum_{k=0}^{n}\boldsymbol{s}(n,k;r,1)D^k, \tag{27}$$

where $\boldsymbol{s}(n,0;r,1)=\delta_{n,0}$ and $\boldsymbol{s}(n,k;r,1)=0$ for $k>n.$

Corollary 3.5

The numbers $\boldsymbol{s}(n,k;r,1)$ satisfy the triangular recurrence relation

$$\boldsymbol{s}(n,k;r,1)=\boldsymbol{s}(n-1,k-1;r,1)+(2n-1)r\boldsymbol{s}(n-1,k;r,1). \tag{28}$$

Proof

The proof follows easily from (22) by setting $s=1$.

Corollary 3.6

The numbers $\boldsymbol{s}(n,k;r,1)$ have the following explicit formula

$$\boldsymbol{s}(n,k;r,1)=r^{n-k}\sum_{\sigma_n=n-k,i_j\in\{0,1\}}\prod_{j=1}^{n}(2j-1)^{i_j}. \tag{29}$$

Proof

The proof follows from (22) by setting $s=1$.

Also, using the recurrence relation (28) we can find the following explicit formula.

Theorem 3.1

The numbers $\boldsymbol{s}(n,k;r,1)$ have the following explicit expression

$$\boldsymbol{s}(n,k;r,1)=\sum_{i_1+\cdots+i_n=k,i_j\in\{0,1\}}\binom{i_1+r}{1-i_1}\binom{i_2+3r}{1-i_2}\cdots\binom{i_n+(2n-1)r}{1-i_n}. \tag{30}$$

Proof

For $k=0$, $\boldsymbol{s}(n,0;r,1)=(r)(3r)(5r)\cdots((2n-1)r)=r^n(1.2.3\cdots(2n-1))$.

For $i_n\in\{0,1\}$, we get

$$\begin{aligned}
&\boldsymbol{s}(n,k;r,1)\\
&=\sum_{i_1+\cdots+i_{n-1}=k-0,i_j\in\{0,1\}}\binom{i_1+r}{1-i_1}\binom{i_2+3r}{1-i_2}\cdots\binom{i_{n-1}+(2n-3)r}{1-i_{n-1}}\cdot(2n-1)r\\
&\quad+\sum_{i_1+\cdots+i_{n-1}=k-1,i_j\in\{0,1\}}\binom{i_1+r}{1-i_1}\binom{i_2+3r}{1-i_2}\cdots\binom{i_{n-1}+(2n-3)r}{1-i_{n-1}}\cdot 1\\
&=\boldsymbol{s}(n-1,k-1;r,1)+(2n-1)r\boldsymbol{s}(n-1,k;r,1).
\end{aligned}$$

That is the same recurrence relation (28) for the numbers $\boldsymbol{s}(n,k;r,1)$. This completes the proof.

iii) If $r=1$ and $s=1$, then we have

Definition 3.4

The numbers $\boldsymbol{s}(n,k):=\boldsymbol{s}(n,k;1,1)$ are defined by

$$\left(\mathrm{e}^{x}D\mathrm{e}^{x}\right)^{n}=\mathrm{e}^{2nx}\sum_{k=0}^{n}\boldsymbol{s}(n,k)D^{k}, \tag{31}$$

where $\boldsymbol{s}(n,0)=\delta_{n,0}$ and $\boldsymbol{s}(n,k)=0$ for $k>n.$

Equation (31) is equivalent to

$$\left(\mathrm{e}^{a^{+}}D\mathrm{e}^{a^{+}}\right)^{n}=\mathrm{e}^{2na^{+}}\sum_{k=0}^{n}\boldsymbol{s}(n,k)a^{k}. \tag{32}$$

Corollary 3.7

The numbers $\boldsymbol{s}(n,k)$ satisfy the triangular recurrence relation

$$\boldsymbol{s}(n+1,k)=\boldsymbol{s}(n,k-1)+(2n+1)\boldsymbol{s}(n,k). \tag{33}$$

Proof

The proof follows by setting $r=1$ in equation (28).

Corollary 3.8

The numbers $\boldsymbol{s}(n,k)$ have the explicit formula

$$s(n,k)=\sum_{\sigma_n=n-k,i_j\in\{0,1\}}\prod_{j=1}^{n}(2j-1)^{i_j}. \tag{34}$$

Proof

The proof follows by setting $r=1$ in equation (29).

Moreover $s(n,k)$ have the following explicit formula.

Corollary 3.9

The numbers $s(n,k)$ have the following explicit expression

$$s(n,k)=\sum_{i_1+\cdots+i_n=k,i_j\in\{0,1\}}\binom{i_1+1}{1-i_1}\binom{i_2+3}{1-i_2}\binom{i_2+5}{1-i_3}\cdots\binom{i_n+(2n-1)}{1-i_n}. \tag{35}$$

Proof

The proof follows by setting $r=1$ in (30).

From equations (29) and (30) (also from equations (34) and (35)) we have the combinatorial identities

$$\sum_{i_1+\cdots+i_n=k,i_j\in\{0,1\}}\binom{i_1+r}{1-i_1}\binom{i_2+3r}{1-i_2}\cdots\binom{i_n+(2n-1)r}{1-i_n}=r^{n-k}\sum_{\sigma_n=n-k,i_j\in\{0,1\}}\prod_{j=1}^{n}(2j-1)^{i_j}. \tag{36}$$

$$\sum_{i_1+\cdots+i_n=k,i_j\in\{0,1\}}\binom{i_1+1}{1-i_1}\binom{i_2+3}{1-i_2}\cdots\binom{i_n+(2n-1)}{1-i_n}=\sum_{\sigma_n=n-k,i_j\in\{0,1\}}\prod_{j=1}^{n}(2j-1)^{i_j}. \tag{37}$$

From equations (29) and (34) we obtain that

$$\boldsymbol{s}(n,k;r,1)=r^{n-k}\boldsymbol{s}(n,k). \tag{38}$$

Remark 3.1

Operating with both sides of equation (13) on the exponential function e^{lx}, we get

$$(l+r_1)^{s_1}(l+2r_1+r_2)^{s_2}+\cdots+(l+2r_1+\cdots+2r_{n-1}+r_n)^{s_n}=\sum_{k=s}^{\beta_n}\boldsymbol{s}(k;\overline{r},\overline{s})l^k.$$

Therefore, since a nonzero polynomial can have only a finite set of zeros, we have

$$\prod_{j=0}^{n-1}\left(2\sum_{m=0}^{j}r_m+r_{j+1}+x\right)^{s_{j+1}}=\sum_{k=0}^{\beta_n}\boldsymbol{s}(k;\overline{r},\overline{s})x^k, r_0=0. \tag{39}$$

If $x=1$, we obtain

$$\sum_{k=0}^{\beta_n}\boldsymbol{s}(k;\overline{r},\overline{s})=\prod_{j=0}^{n-1}\left(1+2\sum_{m=0}^{j}r_m+r_{j+1}\right)^{s_{j+1}}. \tag{40}$$

Remark 3.2

From relation (39), by replacing s_j with s_{j-1}, and relation (18) we conclude that

$$\boldsymbol{s}(k;\overline{r},\tilde{s})=s_{\overline{\alpha}}(n,k;\tilde{s}), \text{where } \alpha_i=-\left(2\sum_{m=0}^{i}r_m+r_{i+1}\right), i=0,1,\cdots,n-1. \tag{41}$$

This gives us a connection between $\mathbf{s}(k;\bar{r},\bar{s})$ and $s_{\bar{a}}(n,k;\tilde{s})$, the generalized Comtet numbers, see [6].

Setting $r_i = r$ and $s_i = s, i = 1,\cdots,n,$ in (39), we get

$$\prod_{j=0}^{n-1}\left((2j+1)r+x\right)^s = \sum_{k=0}^{ns}\mathbf{s}(n,k;r,s)x^k, \tag{42}$$

hence, we have $\mathbf{s}(n,k;r,s) = s_{\bar{\alpha}}(n,k;s),$ where $\alpha_i = -(2i+1)r, i = 0,1,\cdots,n-1,$ see [6].

If $x = 1$, then

$$\sum_{k=0}^{ns}\mathbf{s}(n,k;r,s) = \prod_{j=0}^{n-1}\left((2j+1)r+1\right)^s, n \geq 1. \tag{43}$$

Next we discuss the following special cases of (42) and (43):

I. If $r = 1$, then

II.
$$\left.\begin{aligned} &\prod_{j=0}^{n-1}\left((2j+1)+x\right)^s = \sum_{k=0}^{ns}\mathbf{s}(n,k;1,s)x^k, \\ &\sum_{k=0}^{ns}\mathbf{s}(n,k;1,s) = \prod_{j=0}^{n-1}\left((2j+1)+1\right)^s, n \geq 1, \end{aligned}\right\} \tag{44}$$

hence we have $\mathbf{s}(n,k;1,s) = s_{\bar{\alpha}}(n,k),$ the generalized Comtet numbers, where $\alpha_i = -(2i+1),$

$i = 0,1,\cdots,n-1,$ see [6].

If $s=1$, then we have

$$\left.\begin{aligned}\prod_{j=0}^{n-1}\left((2j+1)r+x\right)=\sum_{k=0}^{n}\boldsymbol{s}(n,k;r,1)x^{k},\\ \sum_{k=0}^{n}\boldsymbol{s}(n,k;r,1)=\prod_{j=0}^{n-1}\left((2j+1)r+1\right),n\geq 1.\end{aligned}\right\} \tag{45}$$

hence we obtain $\boldsymbol{s}(n,k;r,s)=s_{\bar{\alpha}}(n,k)$, Comtet numbers, where $\alpha_i=-(2i+1)r, i=0,1,\cdots,n-1$, see [3] and [4] .

For example if $n=3, r=2$ and s = 2 in (43) we have

$$\sum_{k=0}^{6}\boldsymbol{s}(3,k;2,2)=\prod_{j=0}^{2}(4j+3)^{2}. \tag{46}$$

Using Table 2,

L.H.S. of (46) = s(3,0;2,2) + s(3,1;2,2) + s(3,2;2,2) + s(3,3;2,2) + s(3,4;2,2) + s(3,5;2,2) + s(3,6;2,2) = 14400 + 22080 +12784 + 3552 + 508 + 36 + 1 = 53361.

R.H.S. of (46) = $\prod_{j=0}^{2}(4j+3)^{2}=(3^{2})(7^{2})(11^{2})=53361$.

This confirms (46) and hence (43).

Another example if n = 2, r = 2 and s = 3 in (43) we have

$$\sum_{k=0}^{6}\boldsymbol{s}(2,k;2,3)=\prod_{j=0}^{1}(4j+3)^{3}. \tag{47}$$

Using Table 3,

L.H.S. of (47) = s(2,0;2,3) + s(2,1;2,3) + s(2,2;2,3) + s(2,3;2,3) + s(2,4;2,3) + s(2,5;2,3) + s(2,6;2,3) = 1728 + 3456 + 2736 + 1088 + 228 + 24 + 1 = 9261.

R.H.S. of (46) = $\prod_{j=0}^{1}(4j+3)^3=(3^3)(7^3)=9261$.

This confirms (43).

iii) If $r=s=1$, then we get

$$\left.\begin{aligned}&\prod_{j=0}^{n-1}\left((2j+1)+x\right)=\sum_{k=0}^{n}\boldsymbol{s}(n,k)x^k,\\&\sum_{k=0}^{n}\boldsymbol{s}(n,k)=\prod_{j=0}^{n-1}2(j+1)=2^n n!,n\geq 1,\end{aligned}\right\}\tag{48}$$

hence we have $\boldsymbol{s}(n,k)=s_{\bar{a}}(n,k),$ which is a special case of Comtet numbers, where

$\alpha_i=-(2i+1),i=0,1,\cdots,n-1,$ see [3] and [4] and Table 1.

Setting $e^x=t$, we have $D:=d/dx=t(d/dt)=tD_t=\delta_t,$ then substituting in (2.1) it becomes

$$\left(t^{r_n}\delta_t^{s_n}t^{r_n}\right)\cdots\left(t^{r_2}\delta_t^{s_2}t^{r_2}\right)\left(t^{r_1}\delta_t^{s_1}t^{r_1}\right)=e^{\left(2\sum_{i=1}^{n}r_i\right)x}\sum_{k=0}^{\beta_n}\boldsymbol{s}(k;\bar{r},\bar{s})\delta_t^k.\tag{49}$$

Using, see [12] ,

$$F(\delta)\left(x^{\alpha}f(x)\right)=x^{\alpha}F(\delta+\alpha)f(x),$$

then equation (49) yields

$$(\delta_t + 2r_1 + \cdots + 2r_{n-1} + r_n)^{s_n} \cdots (\delta_t + 2r_2 + r_1)^{s_2} (\delta_t + r_1)^{s_1} = \sum_{k=0}^{\beta_n} \boldsymbol{s}(k; \bar{\boldsymbol{r}}, \bar{\boldsymbol{s}}) \delta_t^k.$$

(50)

Comparing this equation with Equation (4.1) in [6] , we get

$$\boldsymbol{s}(k; \bar{\boldsymbol{r}}, \tilde{\boldsymbol{s}}) = s_{\bar{\alpha}}(n, k; \tilde{\boldsymbol{s}}), \qquad (51)$$

where $\alpha_i = -\left(2\sum_{m=0}^{i} r_m + r_{i+1}\right), i = 0, 1, \cdots, n-1$ and $s_{\bar{\alpha}}(n, k; \tilde{\boldsymbol{s}})$, are the generalized Comtet numbers of the first kind.

Furthermore, using our notations, it is easy from Equation (4.4) in [6] and (41) to show that

$$S(n, i; \bar{\boldsymbol{\alpha}}, \tilde{\boldsymbol{s}}) = \sum_{k=i}^{|s|} \boldsymbol{s}(n, k; \bar{\boldsymbol{r}}, \tilde{\boldsymbol{s}}) S(k, i), |s| = s_0 + s_1 + \cdots + s_{n-1}, \qquad (52)$$

where $\alpha_i = -\left(2\sum_{m=0}^{i} r_m + r_{i+1}\right), i = 0, 1, \cdots, n-1$ and $S(n,k)$ are the Stirling numbers of the second kind.

Next, we find a connection between $s(n, k; r, s)$ and the generalized harmonic numbers $O_n^{(i)}$ which are defined by, see [13] and [14] ,

$$O_n^{(i)} = \sum_{j=1}^{n} \frac{1}{(2j-1)^i}.$$

From (42), we have

$$\sum_{k=0}^{ns} \boldsymbol{s}(n,k;r,s)x^k$$
$$=\prod_{j=0}^{n-1}((2j+1)r)^s \prod_{j=0}^{n-1}(1+x/(2j+1)r)^s = \prod_{j=0}^{n-1}((2j+1)r)^s \exp\left(\log\left(\prod_{j=0}^{n-1}(1+x/(2j+1)r)^s\right)\right)$$
$$=\prod_{j=0}^{n-1}((2j+1)r)^s \exp\left(\sum_{j=0}^{n-1} s\log(1+x/(2j+1)r)\right) = \prod_{j=0}^{n-1}((2j+1)r)^s \exp\left(\sum_{i=1}^{\infty} s\sum_{j=0}^{n-1}\frac{(-1)^{i+1}}{((2j+1)r)^i}\frac{x^i}{i}\right)$$
$$=\prod_{j=0}^{n-1}((2j+1)r)^s \exp\left(s\sum_{i=1}^{\infty} O_n^{(i)}\frac{(-1)^{i+1}}{r^i i}x^i\right) = \prod_{j=0}^{n-1}((2j+1)r)^s \sum_{l=0}^{\infty}\left(s\sum_{i=1}^{\infty} O_n^{(i)}\frac{(-1)^{i+1}}{r^i i}x^i\right)^l \Big/ l!$$
$$=\prod_{j=0}^{n-1}((2j+1)r)^s \sum_{l=0}^{\infty}\frac{s^l}{l!}\sum_{k=l}^{\infty}\sum_{i_1+\cdots+i_l=k}(-1)^{k+l}\frac{O_n^{(i_1)}\cdots O_n^{(i_l)}}{i_1\cdots i_l r^k}x^k$$
$$=\prod_{j=0}^{n-1}((2j+1)r)^s \sum_{l=0}^{\infty}\sum_{k=l}^{\infty}\frac{s^l}{l!}\sum_{i_1+\cdots+i_l=k}(-1)^{k+l}\frac{O_n^{(i_1)}\cdots O_n^{(i_l)}}{i_1\cdots i_l r^k}x^k.$$

Equating the coefficients of x^k on both sides, we obtain

$$\boldsymbol{s}(n,k;r,s) = r^{ns-k}\prod_{j=0}^{n-1}(2j+1)^s \sum_{l=0}^{\infty}\frac{s^l}{l!}\sum_{k=l}^{\infty}\sum_{i_1+\cdots+i_l=k}(-1)^{k+l}\frac{O_n^{(i_1)}\cdots O_n^{(i_l)}}{i_1\cdots i_l r^k}. \tag{53}$$

From (22) and (53), we have the combinatorial identity

$$\sum_{\substack{\sigma_n=ns-k,\\ i_j\geq 0}}\left[\prod_{j=1}^{n}\binom{s}{i_j}(2j-1)^{i_j}\right] = \prod_{j=0}^{n-1}(2j+1)^s \sum_{l=0}^{\infty}\frac{s^l}{l!}\sum_{k=l}^{\infty}\sum_{i_1+\cdots+i_l=k}(-1)^{k+l}\frac{O_n^{(i_1)}\cdots O_n^{(i_l)}}{i_1\cdots i_l r^k}. \tag{54}$$

hence, setting $s=1$, we get the identity

$$\sum_{\substack{\sigma_n=n-k,\\ i_j\geq 0}}\left[\prod_{j=1}^{n}(2j-1)^{i_j}\right] = \prod_{j=0}^{n-1}(2j+1)\sum_{l=0}^{k}\sum_{i_1+\cdots+i_l=k}(-1)^{k+l}\frac{O_n^{(i_1)}\cdots O_n^{(i_l)}}{i_1\cdots i_l l!}. \tag{55}$$

4. SOME APPLICATIONS

4.1. Coherent state and normal ordering

Coherent states play an important role in quantum mechanics especially in optics. The normally ordered form of the boson operator in which all the creation operators a^+ stand to the left of the annihilation operators . Using the properties of coherent states we can define and represent the generalized polynomial $\overline{P}_{r,s}(x)$ and generalized number $\overline{P}_{r,s}$ as follows.

Definition 4.1

The generalized polynomial $\overline{P}_{r,s}(x)$ is defined by

$$\overline{P}_{r,s}(x)=\sum_{k=0}^{\beta_n}\boldsymbol{s}(k;\overline{\boldsymbol{r}},\overline{\boldsymbol{s}})x^k, \tag{56}$$

and the generalized number $\overline{P}_{r,s}$

$$\overline{P}_{r,s}=\overline{P}_{r,s}(1)=\sum_{k=0}^{\beta_n}\boldsymbol{s}(k;\overline{\boldsymbol{r}},\overline{\boldsymbol{s}}). \tag{57}$$

For convenience we apply the convention

$$\boldsymbol{s}(k;\overline{\boldsymbol{r}},\overline{\boldsymbol{s}})=0 \text{ for } k<0 \text{ or } k>\beta_n. \tag{58}$$

Now we come back to normal ordering. Using the properties of coherent states, see [7] , the coherent state matrix element of the boson string yields the generalized polynomial $\overline{P}_{r,s}(x)$

$$\left\langle z\left|\left(e^{r_n a^+} a^{s_n} e^{r_n a^+}\right)\cdots\left(e^{r_2 a^+} a^{s_2} e^{r_2 a^+}\right)\left(e^{r_1 a^+} a^{s_1} e^{r_1 a^+}\right)\right|z\right\rangle$$
$$= \left\langle z\left|e^{\left(2\sum_{l=1}^{n} r_l\right)a^+}\sum_{k=0}^{\beta_n}\boldsymbol{s}(k;\overline{\boldsymbol{r}},\overline{\boldsymbol{s}})a^k\right|z\right\rangle = e^{\left(2\sum_{l=1}^{n} r_l\right)z^*}\sum_{k=0}^{\beta_n}\boldsymbol{s}(k;\overline{\boldsymbol{r}},\overline{\boldsymbol{s}})z^k\langle z|z\rangle = e^{\left(2\sum_{l=1}^{n} r_l\right)z^*}\sum_{k=0}^{n}\boldsymbol{s}(k,\overline{\boldsymbol{r}},\overline{\boldsymbol{s}})z^k$$
$$= e^{\left(2\sum_{l=1}^{n} r_l\right)z^*}\overline{P}_{r,s}(z) = e^{\left(2\sum_{l=1}^{n} r_l\right)z^*}\overline{P}_{r,s}\left(\frac{|z|^2}{z^*}\right), \text{ where } \beta_n = \sum_{j=1}^{n} s_j. \tag{59}$$

Definition 4.2

We define the polynomial $\overline{P}(n,x)$ as

$$\overline{P}(n,x) = \sum_{k=1}^{n}\boldsymbol{s}(n,k)x^k, \tag{60}$$

and the numbers

$$\overline{P}(n) = \overline{P}(n,1) = \sum_{k=1}^{n}\boldsymbol{s}(n,k). \tag{61}$$

For convenience we apply the conventions

$$\boldsymbol{s}(n,0) = \delta_{n,0} \text{ and } \boldsymbol{s}(n,k) = 0 \text{ for } k > n \text{ and } \overline{P}(0) = \overline{P}(0,x) = 1. \tag{62}$$

Similarly, using the properties of coherent states and (32) we have

$$\left\langle z\left|\left(e^{a^+} a e^{a^+}\right)\right|z\right\rangle = \left\langle z\left|\left(e^{2na^+}\sum_{k=0}^{n}\boldsymbol{s}(n,k)\right)a^k\right|z\right\rangle$$
$$= e^{2nz^*}\sum_{k=0}^{n}\boldsymbol{s}(n,k)z^k\langle z|z\rangle = e^{2nz^*}\sum_{k=0}^{n}\boldsymbol{s}(n,k)z^k$$
$$= e^{2nz^*}\overline{P}(n,z) = e^{2nz^*}\overline{P}\left(n,\frac{|z|^2}{z^*}\right). \tag{63}$$

4.2. Matrix Representation

In this subsection we derive a matrix representation of some results obtained.

Let S_r be $n\times n$ lower triangle matrix, where S_r is the matrix whose entries are the numbers $s(n,k;r,1)$,

i.e. $s_r = \left[s(i,j;r,1)\right]_{i,j\geq 0}$. Furthermore let N_r be an $n\times n$ lower triangle matrix defined by $N_r = \left[e^{2irx}(i,j;r,1)\right]_{i,j\geq 0}$, M_r is a diagonal matrix whose entries of the main diagonal are $e^{2irx}, i=0,1,\cdots,n$,

i.e.

$M_r = \operatorname{diag}\left(e^{0rx}, e^{2rx}, e^{4rx}, \cdots, e^{2nrx}\right)$, $R_r = \left(\left(e^{rx}De^{rx}\right)^0, \left(e^{rx}De^{rx}\right)^1, \cdots, \left(e^{rx}De^{rx}\right)^n\right)^T$

and

$D = \left(D^0, D^1, D^2, \cdots, D^n\right)^T$.

Equation (27), may be represented in a matrix form as

$$R_r = N_r D = M_r s_r D, \tag{64}$$

for example if n = 3 then

$$\begin{bmatrix} \left(e^{rx}De^{rx}\right)^0 \\ \left(e^{rx}De^{rx}\right)^1 \\ \left(e^{rx}De^{rx}\right)^2 \\ \left(e^{rx}De^{rx}\right)^3 \end{bmatrix} = \begin{bmatrix} e^{0rx} & 0 & 0 & 0 \\ re^{2rx} & e^{2rx} & 0 & 0 \\ 3r^2e^{4rx} & 4re^{4rx} & e^{4rx} & 0 \\ 15r^3e^{6rx} & 23r^2e^{6rx} & 9re^{6rx} & e^{6rx} \end{bmatrix} \begin{bmatrix} D^0 \\ D^1 \\ D^2 \\ D^3 \end{bmatrix}$$

$$= \begin{bmatrix} e^{0rx} & 0 & 0 & 0 \\ 0 & e^{2rx} & 0 & 0 \\ 0 & 0 & e^{4rx} & 0 \\ 0 & 0 & 0 & e^{6rx} \end{bmatrix} \begin{bmatrix} 1 & 0 & 0 & 0 \\ r & 1 & 0 & 0 \\ 3r^2 & 4r & 1 & 0 \\ 15r^3 & 23r^2 & 9r & 1 \end{bmatrix} \begin{bmatrix} D^0 \\ D^1 \\ D^2 \\ D^3 \end{bmatrix}, \tag{65}$$

its inverse is given by

$$\boldsymbol{D} = \boldsymbol{N}_r^{-1}\boldsymbol{R}_r = \boldsymbol{s}_r^{-1}\boldsymbol{M}_r^{-1}\boldsymbol{R}_r. \tag{66}$$

Setting r = 1 in (64), we get

$$\boldsymbol{R}_1 = \boldsymbol{N}_1\boldsymbol{D} = \boldsymbol{M}_1\boldsymbol{s}_1\boldsymbol{D}, \tag{67}$$

$$\begin{bmatrix} \left(e^{x}De^{x}\right)^0 \\ \left(e^{x}De^{x}\right)^1 \\ \left(e^{x}De^{x}\right)^2 \\ \left(e^{x}De^{x}\right)^3 \end{bmatrix} = \begin{bmatrix} e^{0x} & 0 & 0 & 0 \\ e^{2x} & e^{2x} & 0 & 0 \\ 3e^{4x} & 4e^{4x} & e^{4x} & 0 \\ 15e^{6x} & 23e^{6x} & 9e^{6x} & e^{6x} \end{bmatrix} \begin{bmatrix} D^0 \\ D^1 \\ D^2 \\ D^3 \end{bmatrix} = \begin{bmatrix} e^{0x} & 0 & 0 & 0 \\ 0 & e^{2x} & 0 & 0 \\ 0 & 0 & e^{4x} & 0 \\ 0 & 0 & 0 & e^{6x} \end{bmatrix} \begin{bmatrix} 1 & 0 & 0 & 0 \\ 1 & 1 & 0 & 0 \\ 3 & 4 & 1 & 0 \\ 15 & 23 & 9 & 1 \end{bmatrix} \begin{bmatrix} D^0 \\ D^1 \\ D^2 \\ D^3 \end{bmatrix}, \tag{68}$$

Hence

$$\boldsymbol{D} = \boldsymbol{N}_1^{-1}\boldsymbol{R}_1 = \boldsymbol{s}_1^{-1}\boldsymbol{M}_1^{-1}\boldsymbol{R}_1.$$

For n = 3, we have

$$\begin{bmatrix} D^0 \\ D^1 \\ D^2 \\ D^3 \end{bmatrix} = \begin{bmatrix} 1 & 0 & 0 & 0 \\ -1 & e^{-2x} & 0 & 0 \\ 1 & -4e^{-2x} & e^{-4x} & 0 \\ -1 & 13e^{-2x} & -9e^{-4x} & e^{-6x} \end{bmatrix} \begin{bmatrix} \left(e^x De^x\right)^0 \\ \left(e^x De^x\right)^1 \\ \left(e^x De^x\right)^2 \\ \left(e^x De^x\right)^3 \end{bmatrix} = \begin{bmatrix} 1 & 0 & 0 & 0 \\ -1 & 1 & 0 & 0 \\ 1 & -4 & 1 & 0 \\ -1 & 13 & -9 & 1 \end{bmatrix} \begin{bmatrix} e^{0x} & 0 & 0 & 0 \\ 0 & e^{-2x} & 0 & 0 \\ 0 & 0 & e^{-4x} & 0 \\ 0 & 0 & 0 & e^{-6x} \end{bmatrix} \begin{bmatrix} \left(e^x De^x\right)^0 \\ \left(e^x De^x\right)^1 \\ \left(e^x De^x\right)^2 \\ \left(e^x De^x\right)^3 \end{bmatrix} \tag{69}$$

5. CONCLUSION

In this article we investigated a new family of generalized Stirling numbers of the first kind. Recurrence relations and an explicit formula of these numbers are derived. Moreover some interesting special cases and new combinatorial identities are obtained. A connection between this family and the generalized harmonic numbers is given. Finally, some applications in coherent states and matrix representation of some results are obtained.

REFERENCES

1. El-Desouky, B.S. (2011) Generalized String Numbers of the First Kind: Modified Approach. Journal of Pure and Mathematics: Advances and Applications, 5, 43-59.

2. Gould, H.W. (1964) The Operator and Stirling Numbers of the First Kind. The American Mathematical Monthly, 71, 850-858. >http://html.scirp.org/file/20-7402152x229.png" class="200" /> and Stirling Numbers of the First Kind. The American Mathematical Monthly, 71, 850-858.

3. Comtet, L. (1972) Nombres de stirling généraux et fonctions symétriques. Comptes Rendus de l'Académie des Sciences (Series A), 275, 747-750.

4. Comtet, L. (1974) Advanced Combinatorics: The Art of Finite and Infinite Expiations. D. Reidel Publishing Company, Dordrecht, Holand.

5. Comtet, L. (1973) Une formule explicite pour les puissances successive de l'operateur derivation de Lie. Comptes Rendus de l'Académie des Sciences (Series A), 276, 165-168.

6. El-Desouky, B.S. and Cakić, N.P. (2011) Generalized Higher Order Stirling Numbers. Mathematical and Computer Modelling, 54, 2848-2857.

7. Blasiak, P. (2005) Combinatorics of Boson Normal Ordering and Some Applications. PhD Thesis, University of Paris, Paris.

8. Blasiak, P., Penson, K.A. and Solomon, A.I. (2003) The General Boson Normal Ordering Problem. Physics Letters A, 309, 198-205.

9. Cakic, N.P. (1980) On Some Combinatorial Identities. Applicable Analysis and Discrete Mathematics, 678-715, 91-94.

10. Carlitz, L. (1932) On Arrays of Numbers. American Journal of Mathematics, 54, 739-752.

11. El-Desouky, B.S., Cakic, N.P. and Mansour, T. (2010) Modified Approach to Generalized Stirling Numbers via Differential Operators. Applied Mathematics Letters, 23, 115-120.

12. Viskov, O.V. and Srivastava, H.M. (1994) New Approaches to Certain Identities Involving Differential Operators. Journal of Mathematical Analysis and Applications, 186, 1-10.

13. Macdonald, I.G. (1979) Symmetric Functions and Hall Polynomials. Clarendon (Oxford University) Press, Oxford, London and New York.

14. Choi, J. and Srivastava, H.M. (2011) Some Summation Formulas Involving Harmonic Numbers and Generalized Harmonic Numbers. Mathematical and Computer Modelling, 54, 2220-2234.

CHAPTER 7

Special Numbers on Analytic Functions

Yilmaz Simsek

Department of Mathematics, Faculty of Science,
University of Akdeniz, Antalya, Turkey

ABSTRACT

The aim of this paper is to give some analytic functions which are related to the generating functions for the central factorial numbers. By using these functions and p-adic Volkenborn integral, we derive many new identities associated with the Bernoulli and Euler numbers, the central factorial numbers and the Stirling numbers. We also give some remarks and comments on these analytic functions, which are related to the generating functions for the special numbers.

KEYWORDS

Bernoulli Numbers, Euler Numbers, The Central Factorial Numbers , Array Polynomials, Stirling Numbers of the First Kind and the Second Kind, Generating Function, Functional Equation, Analytic Functions

1. INTRODUCTION, DEFINITIONS AND PRELIMINARIES

The generating functions have been used in many branches of Mathematics and Mathematical Physics. By using p-adic Volkenborn integral, analytic functions and these generating functions, we derive many functional equations and differential equations. By using these equations, we derive many new identities and combinatorics relations related to these numbers and polynomials.

The Stirling numbers of the first and the second are used in pure and applied Mathematics. The Stirling number of the second kind, denoted by $S(n,k)$, is the number of ways to partition a set of n objects into k groups. These numbers occur in combinatorics and in the theory of partitions. The Stirling numbers of the second kind $S(n,v)$ are defined by means of the following generating function:

$$F_S(t,v)=\frac{\left(e^t-1\right)^v}{v!}=\sum_{n=0}^{\infty}S(n,v)\frac{t^n}{n!},$$

(cf. [1] -[19]). These numbers are computed by the following formula:

$$x^n=\sum_{v=0}^{n}\binom{x}{v}v!S(n,v)$$

or

$$S(n,v)=\frac{1}{v!}\sum_{j=0}^{v}\binom{v}{j}(-1)^j(v-j)^n.$$

From the above equation, one has

$$S(n,k)=S(n-1,k-1)+kS(n-1,k),$$

With

$$S(n,0)=0(n\in\square);S(n,n)=1(n\in\square);S(n,1)=1(n\in\square)$$

and $S(n,k)=0$ ($n<k$ or $k<0$) (cf. [1] -[19]).

The Stirling numbers of the first kind $s(n,k)$ are defined by means of the following generating function:

$$\frac{\left(\log(1+x)\right)^v}{v!}=\sum_{n=0}^{\infty}s(n,v)\frac{t^n}{n!},$$

(cf. [19]). These numbers are computed by the following formula:

$$x(x-1)\cdots(x-n+1)=\sum_{v=0}^{n}s(n,v)x^{v}.$$

From the above equation, one has

$$s(n,k)=s(n-1,k-1)-(n-1)s(n-1,k),$$

with

$$s(n,0)=0(n\in\square);s(n,n)=1(n\in\square_{0});s(n,1)=(-1)^{n-1}(n-1)!(n\in\square)$$

and $s(n,k)=0$ ($n<k$ or $k<0$) (cf. [19]).

The array polynomials $S_{v}^{n}(x)$ are defined by means of the following generating function:

$$\frac{\left(e^{t}-1\right)^{v}}{v!}e^{tx}=\sum_{n=0}^{\infty}S_{v}^{n}(x)\frac{t^{n}}{n!},$$

(cf. [1] [13]).

From the above generating function, one has

$$S_{v}^{n}(x)=\frac{1}{v!}\sum_{j=0}^{v}(-1)^{v-j}\binom{v}{j}(x+j)^{n}$$

With

$$S_{0}^{0}(x)=S_{n}^{n}(x)=1,S_{0}^{n}(x)=x^{n}$$

and for $\nu > n$, $S_\nu^n(x) = 0$ (cf. [1] [13]).

2. IDENTITIES AND RELATIONS ON THE CENTRAL FACTORIAL NUMBERS $T(n,k)$

In this section, we study on generating functions for the central factorial numbers $T(n,k)$ on analytic functions. By using these functions, we derive are some functional equations. Applying these functions and equations, we give some identities and relations related to the central factorial numbers $T(n,k)$, which are defined by means of the following generating function (cf. [3] [19]):

$$F_T(t,k) = \frac{1}{(2k)!}\left(e^t + e^{-t} - 2\right)^k = \sum_{n=k}^{\infty} T(n,k)\frac{t^{2n}}{(2n)!}. \tag{2.1}$$

Srivastava and Liu [19] gave many properties and applications of the central factorial numbers.

In the work of Srivastava and Liu [19] , one has the following relations:

$$x^n = \sum_{k=0}^{n} T(n,k)x(x-1)\left(x-2^2\right)\left(x-3^2\right)\cdots\left(x-(k-1)^2\right).$$

Combining the above equation with (1), we also have

$$T(n,k) = T(n-1,k-1) + k^2 T(n-1,k),$$

where $n \geq 1$, $k \geq 1$, $(n,k) \neq (1,1)$.

For n, $k \in$ □, $T(0,k) = T(n,0) = 0$ and $T(n,1) = 1$ (cf. [19]).

In [2] , Cigler gave the following formula for the numbers $T(n,k)$:

$$T(n,k) = \frac{1}{(2k)!} \sum_{j=0}^{2k} (-1)^j \binom{2k}{j} (j-k)^{2n}.$$

Some properties of the function $F_T(t,k)$ are given as follows:

$$F_T(0,k) = 1.$$

From this property, one has $T(0,0) = 1$. The function $F_T(t,k)$ is related to the hyperbolic functions and generating functions for the special numbers. Thus we get the following functional equation:

$$F_T(t,k) = \frac{2^k}{(2k)!} (\cosh(t) - 1)^k$$

The function $F_T(t,k)$ is related to the generating function for the Stirling numbers of the second kind, $S(n,k)$. That is

$$F_T(t,k) = F_S(t,2k) e^{-kt}, \tag{2.2}$$

Where

$$F_S(t,2k) = \frac{1}{(2k)!} (e^t - 1)^{2k} = \sum_{n=0}^{\infty} S(n,2k) \frac{t^n}{n!}.$$

And

$$F_T(t,k)=\frac{k!}{(2k)!}\sum_{j=0}^{k}F_S(t,j)F_S(-t,k-j). \tag{2.3}$$

We are going to differentiate (1) with respect to t to derive a recurrence relation or identities for the central factorial numbers.

$$\frac{d}{dt}F_T(t,k)=\frac{k(2k-2)!}{(2k)!}\left(e^t-e^{-t}\right)F_T(t,k-1) \tag{2.4}$$

or

$$\frac{d}{dt}F_T(t,k)=\frac{2k(2k-2)!}{(2k)!}F_T(t,k-1)\sinh(t). \tag{2.5}$$

Lemma 1 ([9] , Lemma 11, Equation (7)]

$$\sum_{n=0}^{\infty}\sum_{k=0}^{\infty}A(n,k)=\sum_{n=0}^{\infty}\sum_{k=0}^{\left[\frac{n}{2}\right]}A(n,n-2k),$$

where $[x]$ denotes the greatest integer function.

Remark 1 Lemma 1 was generalized by Srivastava et al. [16] . By using these generalization, one may generalize Theorem 1.

Theorem 1

$$S(n,2k)=n!\sum_{j=0}^{\left[\frac{n}{2}\right]}\frac{k^{n-2j}}{(2j)!(n-2j)!}T(j,k).$$

Proof. By combining (2) with (1), we get

$$\sum_{n=0}^{\infty} S(n,2k)\frac{t^n}{n!} = \sum_{n=0}^{\infty}\frac{(kt)^n}{n!}\sum_{n=0}^{\infty} T(n,k)\frac{t^{2n}}{(2n)!}.$$

By using Lemma 1, we obtain

$$\sum_{n=0}^{\infty} S(n,2k)\frac{t^n}{n!} = \sum_{n=0}^{\infty}\left(\sum_{j=0}^{\left[\frac{n}{2}\right]}\frac{k^{n-2j}}{(2j)!(n-2j)!}T(j,k)\right)t^n.$$

Comparing the coefficients of t^n on both sides of the above equation, we arrive at the desired result. □

Theorem 2

$$\sum_{j=0}^{\left[\frac{n}{2}\right]}\frac{k^{n-2j}}{(2j)!(n-2j)!}T(j,k) = \frac{k!}{n!(2k)!}\sum_{j=0}^{k}\sum_{l=0}^{n}(-1)^{k-j}\binom{n}{l}S_j^l(j)S(n-l,k-j),$$

where $S_j^l(j)$ denotes the array polynomials.

Proof. By using (3), we derive the following functional equation:

$$F_T(t,k)e^{tx} = \frac{k!}{(2k)!}\sum_{j=0}^{k}(-1)^{k-j}F_A(t,j)F_S(t,k-j),$$

(2.6)

where $F_A(t,j)$ denotes generating function for the array polynomials $S_j^n(x)$. By combining (2.6) with (2.1), we obtain

$$\frac{k!}{(2k)!}\sum_{j=0}^{k}(-1)^{k-j}\sum_{n=0}^{\infty}S_j^n(j)\frac{t^n}{n!}\sum_{n=0}^{\infty}S(n,k-j)\frac{t^n}{n!}=\sum_{n=0}^{\infty}\frac{(kt)^n}{n!}\sum_{n=0}^{\infty}T(n,k)\frac{t^{2n}}{(2n)!}.$$

By using Lemma 1, we obtain

$$\sum_{n=0}^{\infty}\left(\sum_{j=0}^{\left[\frac{n}{2}\right]}\frac{k^{n-2j}}{(2j)!(n-2j)!}T(j,k)\right)t^n=\sum_{n=0}^{\infty}\left(\frac{k!}{n!(2k)!}\sum_{j=0}^{k}\sum_{l=0}^{n}(-1)^{k-j}\binom{n}{l}S_j^l(j)S(n-l,k-j)\right)\frac{t^n}{n!}.$$

Comparing the coefficients of t^n on both sides of the above equation, we arrive at the desired result.

Remark 2 By using functional equation in (2.3), (2.4) and (2.5), one may also obtain an interesting identities and relationships between the Stirling numbers of the second kind and the central factorial numbers.

3. IDENTITIES RELATED TO THE P-ADIC VOLKENBORN INTEGRAL AND THE CENTRAL FACTORIAL NUMBERS

In this section, we give applications of the p-adic integral on $\mathbb{Z}_p$, set of p-adic integers. By using the Witt's formula for the Bernoulli and Euler numbers related to the p-adic Volkenborn integral and the Mahler coefficients, we derive many interesting and novel identities associated with the central factorial numbers, the family of numbers $C_n^{(x)}$, the Stirling numbers of the first and the second kind and also combinatorics identities.

We use notations of the work of Kim [4] for the p-adic q-Volkenborn integral. Let p be a fixed prime. It is known that

$$\mu_q\left(x+p^N\mathbb{Z}_p\right)=\frac{q^x}{\left[p^N\right]_q}$$

is a distribution on $\mathbb{Z}_p$ for $q\in\mathbb{Z}_p$ with $|1-q|_p<1$, cf. [4]. Let $UD(\mathbb{Z}_p)$ be the set of uniformly differentiable functions on $\mathbb{Z}_p$. The p-adic q-integral of the function $f\in UD(\mathbb{Z}_p)$ is defined by Kim [4] as follows:

$$\int_{\mathbb{Z}_p} f(x)\mathrm{d}\mu_q(x)=\lim_{N\to\infty}\frac{1}{\left[p^N\right]_q}\sum_{x=0}^{p^N-1}f(x)q^x,$$

Where

$$[x]=\frac{1-q^x}{1-q}.$$

The bosonic p-adic integral (p-adic Volkenborn integral) was considered from a physical point of view to the bosonic limit $q\to 1$, as follows ([4]):

$$\int_{\mathbb{Z}_p} f(x)\mathrm{d}\mu_1(x)=\lim_{N\to\infty}\frac{1}{p^N}\sum_{x=0}^{p^N-1}f(x), \tag{3.1}$$

Where

$$\mu_1\left(x+p^N\mathbb{Z}_p\right)=\frac{1}{p^N}.$$

The p-adic q-integral is used in many branch of mathematics, mathematical physics and other areas (cf. [4] [11] [14] [18]).

By using (7), we have the Witt's formula for the Bernoulli numbers B_n as follows:

$$\int_{\mathbb{Z}_p} x^n \mathrm{d}\mu_1(x)=B_n \tag{3.2}$$

(cf. [4] [5] [11] [14] [18]).

We consider the fermionic integral in contrast to the bosonic integral, which is called the fermionic p-adic Volkenborn integral on $\mathbb{Z}_p$ cf. [5] . That is

$$\int_{\mathbb{Z}_p} f(x)\mathrm{d}\mu_{-1}(x)=\lim_{N\to\infty}\sum_{x=0}^{p^N-1}(-1)^x f(x) \tag{3.3}$$

Where

$$\mu_1\left(x+p^N\mathbb{Z}_p\right)=\frac{(-1)^x}{p^N}$$

(cf. [5]). By using (9), we have the Witt's formula for the Euler numbers E_n as follows:

$$\int_{\mathbb{Z}_p} x^n \mathrm{d}\mu_{-1}(x) = E_n, \tag{3.4}$$

(cf. [5] [6] [11] [18]).

Theorem 3

$$\int_{\mathbb{Z}_p} \binom{x}{j} \mathrm{d}\mu_1(x) = \frac{(-1)^j}{j+1}. \tag{3.5}$$

Proof of Theorem 3 was given by Schikhof [14] .

Theorem 4

$$\int_{\mathbb{Z}_p} \binom{x}{j} \mathrm{d}\mu_{-1}(x) = \frac{(-1)^j}{2^j}.$$

Proof of Theorem 4 was given by Kim et al. ([7] [8]).

The family of numbers $C_n^{(x)}$ was defined by Srivastava and Liu [19] . These numbers are related to the Stirling numbers of the first and the second kind, which are given by the following theorem.

Theorem 5 ([19] , Theorem 1]) Let $n \geq k \ (n,k \in \mathbb{N})$ and

$$w(n,k) = (-1)^k \sum_{j=k}^{n} 2^{n-j} S(n,j) s(j,k).$$

Then

$$C_n^{(x)} = \sum_{k=1}^{n} w(n,k)x^k. \tag{3.6}$$

In [19] , Equation-(2.3)], Srivastava and Liu also gave the following relation:

$$C_n^{(x)} = \sum_{j=0}^{n} (-1)^j j!2^{n-j} \binom{x+j-1}{j} S(n,j). \tag{3.7}$$

By using the above formulae, we derive the following identities related to the numbers $C_n^{(x)}$, the Stirling numbers, the Bernoulli numbers and the Euler numbers.

According to the work of Kim at al. ([7] [8]), we get

$$\int_{\mathbb{Z}_p} \binom{-x}{j} d\mu_{-1}(x) = \int_{\mathbb{Z}_p} \binom{x+j-1}{j} d\mu_{-1}(x).$$

Combining Theorem 4 with the above equation, we get

$$\int_{\mathbb{Z}_p} \binom{x+n-1}{n} d\mu_{-1}(x) = \sum_{m=0}^{n} (-1)^m \binom{n-1}{n-m} 2^{-m}. \tag{3.8}$$

Integrating both sides of (3.7) with respect to the fermionic integral on $\mathbb{Z}_p$, and than using (3.8), we obtain the following Lemma:

Lemma 2

$$\int_{\mathbb{Z}_p} C_n^{(x)} \mathrm{d}\mu_{-1}(x) = \sum_{j=0}^{n} (-1)^j \, j! S(n,j) \sum_{m=0}^{j} (-1)^m \binom{j-1}{j-m} 2^{n-m-j}.$$

Integrating both sides of (3.6) with respect to the fermionic integral on $\mathbb{Z}_p$, and than using (3.4), we obtain the following Lemma:

Lemma 3

$$\int_{\mathbb{Z}_p} C_n^{(x)} \mathrm{d}\mu_{-1}(x) = \sum_{k=0}^{n} E_k w(n,k),$$

where E_k denotes the Euler numbers.

By using Lemma 2 and Lemma 3, we arrive at the following theorem:

Theorem 6

$$\sum_{k=0}^{n} E_k w(n,k) = \sum_{j=0}^{n} (-1)^j \, j! S(n,j) \sum_{m=0}^{j} (-1)^m \binom{j-1}{j-m} 2^{n-m-j}.$$

Similarly, according to the work of Kim et al. ([7] [8]), we get

$$(-1)^n \int_{\mathbb{Z}_p} \binom{-x}{j} \mathrm{d}\mu_1(x) = \int_{\mathbb{Z}_p} \binom{x+n-1}{n} \mathrm{d}\mu_1(x).$$

Combining Theorem 3 with the above equation, we get

$$\int_{\mathbb{Z}_p}\binom{x+j-1}{j}d\mu_1(x)=\sum_{m=0}^{j}(-1)^m\binom{j-1}{j-m}\frac{1}{m+1}. \tag{3.9}$$

Integrating both sides of (3.7) with respect to the bosonic p-adic integral on $\mathbb{Z}_p$, and than using (3.9), we obtain the following lemma:

Lemma 4

$$\int_{\mathbb{Z}_p}C_n^{(x)}d\mu_1(x)=\sum_{j=0}^{n}(-1)^j j!S(n,j)\sum_{m=0}^{j}(-1)^m\binom{j-1}{j-m}\frac{2^{n-j}}{m+1}.$$

Integrating both sides of (3.6) with respect to the fermionic integral on $\mathbb{Z}_p$, and than using (3.2), we obtain the following lemma:

Lemma 5

$$\int_{\mathbb{Z}_p}C_n^{(x)}d\mu_1(x)=\sum_{k=0}^{n}B_k w(n,k),$$

where B_k denotes the Bernoulli numbers.

By using Lemma 4 and Lemma 5, we arrive at the following theorem:

Theorem 7

$$\sum_{k=0}^{n}B_k w(n,k)=\sum_{j=0}^{n}(-1)^j j!S(n,j)\sum_{m=0}^{j}(-1)^m\binom{j-1}{j-m}\frac{2^{n-j}}{m+1}.$$

ACKNOWLEDGEMENTS

The paper was supported by the Scientific Research Project Administration of Akdeniz University.

REFERENCES

1. Chang, C.-H. and Ha, C.-W. (2006) A Multiplication Theorem for the Lerch Zeta Function and Explicit Representations of the Bernoulli and Euler polynomials. Journal of Mathematical Analysis and Applications, 315, 758-767.

2. Cigler, J. Fibonacci Polynomials and Central Factorial Numbers. Preprint.http://homepage.univie.ac.at/johann.cigler/preprints/central-factorial.pdf

3. Comtet, L. (1974) Advanced Combinatorics: The Art of Finite and Infinite Expansions. Reidel, Dordrecht and Boston, (Translated from the French by J. W. Nienhuys).

4. Kim, T. (2002) q-Volkenborn integration. Russian Journal of Mathematical Physics, 19, 288-299.

5. Kim, T. (2006) q-Euler Numbers and Polynomials Associated with p-Adic q-Integral and Basic q-zeta Function. Trends in International Mathematics and Science Study, 9, 7-12.

6. Jang, L.C. and Kim, T. (2008) A New Approach to q-Euler Numbers and Polynomials. Journal of Concrete and Applicable Mathematics, 6, 159-168.

7. Kim, D.S. and Kim, T. (2013) Daehee Numbers and Polynomials. Applied Mathematical Sciences, 7, 5969-5976.

8. Kim, D.S., Kim, T. and Seo, J. (2013) A Note on Changhee Numbers and Polynomials. Advanced Studies in Theoretical Physics, 7, 993-1003.

9. Rainville, E.D. (1960) Special Functions. The Macmillan Company, New York.

10. Simsek, Y. (2010) On q-Deformed Stirling numbers. International Journal of Computer Mathematics, 15, 70-80.

11. Simsek, Y. (2010) Complete sum of products of (h,q)-Extension of Euler Polynomials and Numbers. Journal of Difference Equations and Applications, 16, 1331-1348.

12. Simsek, Y. (2013) Identities Associated with Generalized Stirling Type Numbers and Eulerian Type Polynomials. Mathematical and Computational Applications, 18, 251-263.

13. Simsek, Y. (2013) Generating Functions for Generalized Stirling type Numbers, Array Type Polynomials, Eulerian Type Polynomials and Their Applications. Fixed Point Theory and Applications, 87, 343-1355.

14. Schikhof, W.H. (1984) Ultrametric Calculus: An Introduction to p-Adic Analysis. Cambridge Studies in Advanced Mathematics 4, Cambridge University Press, Cambridge.

15. Srivastava, H.M. (2011) Some Generalizations and Basic (or q-) Extensions of the Bernoulli, Euler and Genocchi Polynomials. Applied Mathematics & Information Sciences, 5, 390-444.

16. Srivastava, H.M., Ozarslan, M.A. and Kaanoğlu, C. (2010) Some Families of Generating Functions for a Certain Class of Three-Variable Polynomials. Integral Transforms and Special Functions, 21, 885-896.

17. Srivastava, H.M. and Choi, J. (2012) Zeta and q-Zeta Functions and Associated Series and Integrals. Elsevier Science Publishers, Amsterdam, London and New York.

18. Srivastava, H.M., Kim, T. and Simsek, Y. (2005) q-Bernoulli Numbers and Polynomials Associated with Multiple q-Zeta Functions and Basic L-Series. Russian Journal of Mathematical Physics, 12, 241-268.

19. Srivastava, H.M. and Liu, G.-D. (2009) Some Identities and Congruences Involving a Certain Family of Numbers. Russian Journal of Mathematical Physics, 16, 536-542.

CHAPTER 8

A Scale Invariant Distribution of the Prime Numbers

Wayne S. Kendal [1,*,†] and Bent Jørgensen [2,†]

[1]Division of Radiation Oncology, University of Ottawa, 501 Smyth Road, Ottawa, ON K1H 8L6, Canada

[2]Department of Mathematics and Computer Science, University of Southern Denmark, Campusvej 55, Odense M DK-5230, Denmark

ABSTRACT

The irregular distribution of prime numbers amongst the integers has found multiple uses, from engineering applications of cryptography to quantum theory. The degree to which this distribution can be predicted thus has become a subject of current interest. Here, we present a computational analysis of the deviations between the actual positions of the prime numbers and their predicted positions from Riemann's counting formula, focused on the variance function of these deviations from sequential enumerative bins. We show empirically that these deviations can be described by a class of probabilistic models known as the Tweedie exponential dispersion models that are characterized by a power law relationship between the variance and the mean, known by biologists as Taylor's power law and by engineers as fluctuation scaling. This power law behavior of the prime number deviations is remarkable in that the same behavior has been found within the distribution of genes and single nucleotide polymorphisms (SNPs) within the human genome, the distribution of animals and plants within their habitats, as well as within many other biological and physical processes. We explain the common features of this behavior through a statistical convergence effect related to the central limit theorem that also generates $1/f$ noise.

KEYWORDS

exponential dispersion models; $1/f$ noise; fractal scaling; self-organized criticality

1. INTRODUCTION

Prediction of the positions of the prime numbers within the sequence of integers has been a goal that has long interested mathematicians, engineers, and physicists [1]. The prime counting function $\pi(x)$ gives the actual number of prime numbers up to the positive real value x and takes the form of a step function that increases by the value of 1 with each new prime number. In 1808, Legendre empirically showed that $\pi(x)$ could be approximated by the formula $x/[\log(x) - B]$; B being a numerical constant [2]. A further improvement in the estimation of $\pi(x)$ was provided by Gauss, who used the logarithmic integral, defined on the positive real numbers $x \neq 1$ for this purpose [3]:

$$\mathrm{li}(x) = \int_0^x dt / \ln(t) \tag{1}$$

These approximations lead to the prime number theorem:

$$\pi(x) \sim x / \ln(x) \tag{2}$$

with initial proofs delivered by Hadamard [4] and de la Vallée Poussin [5] in 1896. A more detailed formula for the general trend in the position of prime numbers has been attributed to Bernhard Riemann [6]:

$$R(x) = \sum_{n=1}^{\infty} \frac{\mu(n)}{n} \mathrm{li}(x^{1/n}) \tag{3}$$

expressed in terms of the Möbius function on the integer values n:

$$\mu(n) = \begin{cases} 1 & \text{if } n = 1 \\ 0 & \text{if } n \text{ has one or more repeated prime factors} \\ (-1)^k & \text{if } n \text{ is a product of } k \text{ distinct primes} \end{cases} \tag{4}$$

The present report focuses on the local behavior of the deviations between $\pi(x)$ and $R(x)$ that have been conjectured to obey the equation [7]:

$$D(x) = R(x) - \pi(x) \sim \sum_{\rho} R(x^{\rho}) \tag{5}$$

The summation here is specified over the nontrivial (complex) zeros ρ of the Riemann zeta function, providing a link to Riemann's hypothesis that the nontrivial zeros of the Riemann zeta function should have as their real component the value of ½ [6]. An equivalent form of this (yet unproven) hypothesis can be stated using the summatory Mertens function [8]:

$$M(n)=\sum_{i=1}^{n}\mu(i) \tag{6}$$

namely that:

$$M(n)=O(n^{1/2+\varepsilon}) \tag{7}$$

for every positive ε [9].

The deviations *D*(*x*) empirically reveal chaotic features, as well as long range correlations [10], that would seem to indicate an underlying structure. Indeed, the positional irregularities of prime numbers have the characteristics of *1/f* noise [11], and can be related to the eigenvalues of random matrices from the Gaussian unitary ensemble that are employed in quantum chaos [12]. It remains unclear, though, why the distribution of prime numbers should exhibit such features. Here these deviations are empirically shown to correspond to a probabilistic model characterized by a power law relationship between the variance and the mean. This model belongs to the class of scale invariant, or Tweedie, exponential dispersion models [13] which appear as weak limits for other models under a certain central limit-like effect [14]. This convergence effect, we will argue, leads to a wide manifestation of these models within complex natural and physical systems, analogous to what is seen with the Gaussian distribution within more restricted systems governed by the central limit theorem itself. Moreover, this Tweedie convergence effect can explain the genesis of $1/f$ noise in such sequences that has implications regarding self-organized criticality [15].

2. EXPERIMENTAL SECTION

The Riemann prime counting function Equation (3) was estimated here using the Gram series [16]. The absolute values of the prime number deviations $|D(x)|$ were estimated for sequential equal-sized enumerative bins that span the integers. The values of $|D(x)|$ from within each bin were summed and the means and variances of these sums estimated over the region of interest. This process was repeated for successively larger bin sizes so that a log-log plot could be constructed of the sampled variances *versus* the sampled means with different sized bins. A straight line relationship in such a plot would imply a power law relationship between the variance and the mean, variously referred to in the literature as Taylor's ecological power law as well as fluctuation scaling.

Empirical cumulative distribution functions (CDFs) were also constructed from the primary sequences of $|D(x)|$ and then compared to the theoretical CDF from the Tweedie compound Poisson-gamma distribution [13], a distribution which inherently expresses a variance to mean power law with its power law exponent constrained to range between 1 and 2.

3. RESULTS AND DISCUSSION

3.1. The Variance Function of the Prime Number Deviations

We empirically reviewed the absolute values of the deviations $|D(x)|$ for the first 50,000 integers (Figure 1), focusing on their variance function [13]: The mean and variance of the sequence $|D(x)|$ were estimated and, to extend the range of values for the mean, the sequence $|D(x)|$ was divided into non-overlapping and adjacent counting bins, 10 integers long. The values within each bin were summed and the mean and variance of these sums were estimated over all the bins. This process was repeated for successively larger and larger bin sizes. Figure 2a demonstrates the log-log plot of the empirical variances *versus* their respective means. A linear relationship was evident on the log-log plot, indicative of a power law relationship σ2=a·m¯p between the variance σ2 and the mean m¯ with exponent $p = 1.83$.

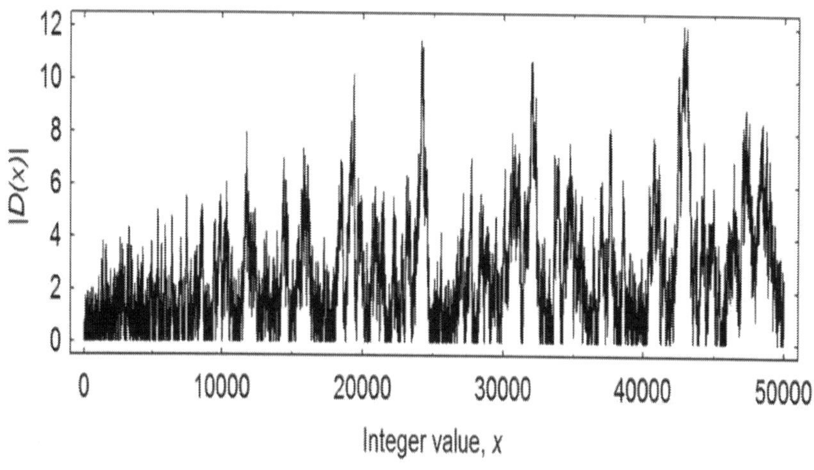

Figure 1. The absolute value $|D(x)|$ of the discrepancy between the prime counting function $\pi(x)$ and Riemann's formula $R(x)$ for the first 50,000 integers.

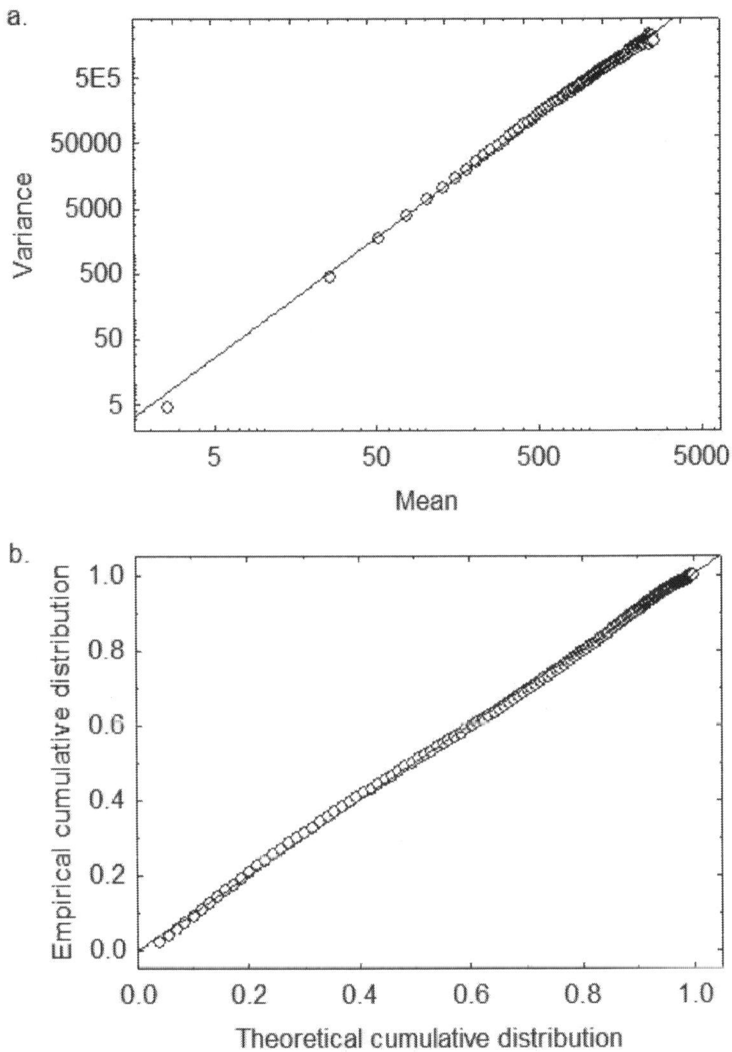

Figure 2. (**a**) The variance function for the deviations of the prime numbers. A power law relationship was obtained with exponent p = 1.83 and constant a = 1.37; (**b**) Probability-probability plot. A frequency histogram was constructed for the values $|D(x)|$. The empirical CDF obeyed a Tweedie compound Poisson-gamma distribution with $\theta = -0.837$, $\lambda = 0.934$, $\alpha = -0.586$ and $p = 1.63$.

We sought to determine the behavior of the power law for the deviations $|D(x)|$ beyond the first 50,000 integers in order to make sure that this behavior was not simply a loco-regional artifact. Figure 3a provides this function determined from the integers $x = 10^6...10^7$, in steps of 10^3. The variance to mean power law fitted this sequence well with $p = 1.66$. In Figure 3c,e this process was repeated for the integer sequences $x = 10^8...10^9$ in steps of 10^5 and $x = 10^{11}...10^{12}$ in steps of 10^8, to reveal that variance to mean power law was evident for over 12 orders of magnitude.

3.2. Prime Number Deviations as a Self-Similar Process

In our analysis we examined the sequence of prime number deviations $Y = (Y_i = |D(i)|: i = 1, 2,..., N)$ using expanding enumerative bins. Let the mean and variance of this initial sequence be represented by the constants $\hat{\mu} = E[Y]$ and $\hat{\sigma}^2 = \mathrm{var}[Y]$ respectively. One can construct a sequence of equal-sized and non-overlapping enumerative bins of integer size m to yield a new set of sequences $Y^{(m)}$ with the reproductive property:

$$Y_i^{(m)} = \frac{1}{m}(Y_{im-m+1} + ... + Y_{im}), \quad i > 1 \tag{8}$$

The values of m were chosen here so that N/m was an integer. We then have E[Y(m)]=E[Y]=μ^, which is a constant provided that the initial sequence remains unaltered.

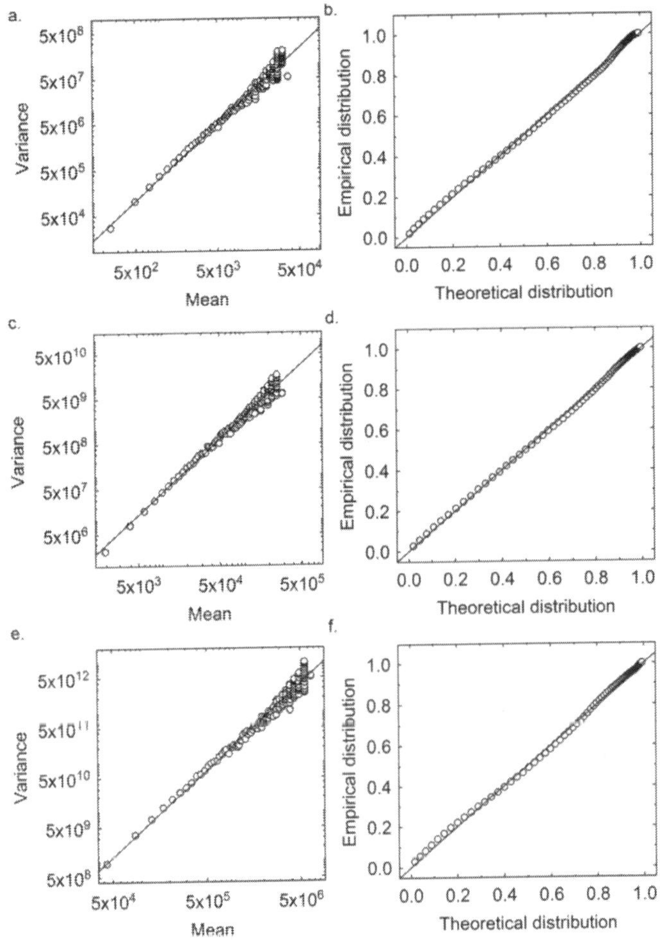

Figure 3. (**a**) Variance function of $|D(x)|$ for $x = 10^6 \ldots 10^7$ in steps of 10^3. Linear regression gave $p = 1.66$ and $a = 3.53$; (**b**) CDF with $x = 10^6 \ldots 10^7$ in steps of 10^3. Here $\theta = -0.096$, $\lambda = 0.62$, $\alpha = -0.38$ and $p = 1.73$; (**c**) Variance function of $|D(x)|$ for $x = 10^8 \ldots 10^9$in steps of 10^5. Here $p = 1.67$ and $a = 9.28$; (**d**) CDF with $x = 10^8 \ldots 10^9$ in steps of 10^5. Here $\theta = -9.94 \times 10^{-3}$, $\lambda = 0.34$, $\alpha = -0.30$ and $p = 1.77$; (**e**) Variance function of $|D(x)|$ for$x = 10^{11} \ldots 10^{12}$ in steps of 10^8. Here $p = 1.74$ and $a = 8.57$; (**f**) CDF with $x = 10^{11} \ldots 10^{12}$ in steps of 10^8. Here $\theta = -5.37 \times 10^{-4}$, $\lambda = 0.067$, $\alpha = -0.41$ and $p = 1.71$.

Alternatively, the enumerative bins can be used to construct a set of additive sequences $Z^{(m)}$ so that:

$$Z_i^{(m)} = (Y_{im-m+1} + \ldots + Y_{im}) \tag{9}$$

as was done with the prime number deviations $|D(x)|$. These additive and reproductive sequences are related to each other by the equation $Z_i^{(m)} = mY_i^{(m)}$. We have for their means and variances, $\mathrm{E}[Z^{(m)}] = m^2 \operatorname{var}[Y^{(m)}]$.

The variance to mean power law, demonstrated with the prime number deviations, implies the existence of long range correlations within the initial sequence of deviations. Long range correlations have been studied by Leland [17] in the context of self-similar processes, where the general form for the correlation function $r(k)$ takes the form,

$$r(k) = k^{-\beta} L(k), \quad k \to \infty \tag{10}$$

Here, k is the autocorrelation shift, β is a real valued constant such that $0 < \beta < 1$, and $L(k)$ is a slowly varying function for large values of k. For large values of k, this correlation approximates a power law with $r(k) \sim k^{-\beta}$.

Consider the case where the reproductive sequences $Y^{(m)}$ obey the property:

$$\operatorname{var}[Y^{(m)}] = \hat{\sigma}^2 m^{-\beta} \tag{11}$$

Tsybakov and Georganis [18] have shown that Equation (11) holds if and only if the autocorrelation for the sequence Y is:

$$r(k) = \frac{1}{2}[(k+1)^{2-\beta} - 2k^{2-\beta} + (k-1)^{2-\beta}] \tag{12}$$

In the limit this formula takes the form [18]:

$$\lim_{k\to\infty} \frac{r(k)}{k^{-\beta}} = \frac{1}{2}(2-\beta)(1-\beta)) \tag{13}$$

where the constant β is related to the Hurst exponent *H*, β = 2 (1− *H*) For additive sequences the variance takes the form:

$$\mathrm{var}[Z_i^{(m)}] = m^2 \,\mathrm{var}[Y^{(m)}] = (\hat{\sigma}^2 / \hat{\mu}^{2-\beta}) E[Z_i^{(m)}]^{2-\beta} \tag{14}$$

Since μ^ and σ^2 are constant, this represents a variance to mean power law with exponent $p = 2 - \beta = 2H$. The power law found from the prime number deviations revealed that $p = 1.83$, and so the Hurst exponent was $H = 0.92$.

3.3. 1/f Noise from the Prime Number Deviations

Autocorrelations of the form $r(k) \sim k{-}\beta$ imply, by virtue of the Wiener-Khintchine theorem [19], that the corresponding power spectral density will approximate $S(f) \sim f^{\beta-1}$, where *f* is the frequency. In the case of the prime number deviations $|D(x)|$, this would imply $S(f) \sim f^{-0.83}$, where the frequency is determined from the spacing of the consecutive values of the data sequence. Power spectra like this, with $S(f) \propto 1/f\gamma$ and $0 < \gamma < 2$, are the hallmark of 1/*f* noise.

3.4. Prime Number Deviations Described by a Tweedie Exponential Dispersion Model

In addition to $1/f$ noise, the variance-to-mean power law is associated with a class of probability distributions, known as the Tweedie exponential dispersion models, named after the man who first described them [20]. There are several types of Tweedie models, each distinguished by the values of their power law exponent p, including the Gaussian ($p = 0$), Poisson ($p = 1$), compound Poisson-gamma (PG) ($1 < p < 2$), gamma ($p = 2$) and inverse Gaussian ($p = 3$) distributions [13]. We employ here the class of PG distributions, characterized by gamma distributed jumps over the domain of the positive real numbers plus zero, to model the deviations $|D(x)|$. The additive form for the cumulant generating function for this PG distribution can be written as:

$$K^*(s) = \lambda\kappa(\theta)\left[\left(1+\frac{s}{\theta}\right)^{\alpha} - 1\right] \tag{15}$$

Here, s is the variable of the generating function, θ is the canonical parameter, λ is the index statistic, and the constant α and the cumulant function $k(\theta)$ are given by:

$$\alpha = (p-2)/(p-1) \text{ and} \tag{16}$$

$$\kappa(\theta) = \frac{\alpha-1}{\alpha}\left(\frac{\theta}{\alpha-1}\right)^{\alpha} \tag{17}$$

The PG distribution is thus specified by 3 independent adjustable parameters α, λ and θ. The parameter α relates to Taylor's power law exponent p, whereas λ and θ relate to the notions of shape and scale parameter employed in the conventional description of distributions.

The probability density function on the variable z, $p^*(z, \theta, \lambda, \alpha)$, that corresponds to this distribution does not exist in closed form. However, it can be expressed as an infinite series [13]:

$$p^*(z;\theta,\lambda,\alpha) = c^*(z;\lambda)\exp[\theta \cdot z - \lambda\kappa(\theta)], \text{ with} \quad (18)$$

$$c^*(z;\lambda) = \begin{cases} \dfrac{1}{z}\sum_{n=1}^{\infty} \lambda^n \kappa^n(-1/z)/\Gamma(-\alpha \cdot n)n! & \text{for } z > 0 \\ 1 & \text{for } z = 0 \end{cases} \quad (19)$$

Figure 2b provides a probability-probability plot of the respective CDF fitted to data derived from the sequence of $|D(x)|$ from the first 50,000 integers. The empirical and theoretical CDFs agreed well. Similarly Figure 3b provides the probability-probability plot corresponding to $|D(x)|$ derived from the integers $x = 10^6 ... 10^7$, in steps of 10^3. The Tweedie PG CDF agreed closely with the empirical CDF. Figure 3d,f provides the probability-probability plots corresponding to the Tweedie PG distribution fitted from the prime number deviations corresponding to the integer sequences $x = 10^8 ... 10^9$ in steps of 10^5 and$x = 10^{11} ... 10^{12}$ in steps of 10^8. The Tweedie PG distribution thus appeared to remain valid for over 12 orders of magnitude of data.

3.5. The Tweedie Convergence Theorem

For exponential dispersion models $ED(\mu, \sigma^2)$ with unit variance functions of the form $V(\mu) \propto \mu p$ Jørgensen *et al.* have shown that as $\mu \to 0$ or $\mu \to \infty$ then, within the constant factor c, $c - 1$ $ED(c\mu, \sigma^2 c^{2-p})$ will converge to the form of a Tweedie model as either $c \to 0$ or $c \to \infty$ [14]. Since the variance functions for many probability distributions approximate the form $V(\mu) \propto \mu p$, for small or large values of μ, the Tweedie distributions will act as the foci of convergence for a wide variety of data [14]. This convergence property appears related to stable generalizations of the CLT [13], suggesting that the Tweedie models have a role analogous to that of the Gaussian distribution in statistical theory. As will be shown below, this theorem has implications with respect to the distribution of prime numbers. Parenthetically, it should also be mentioned that the mean and variance of an additive random variable Z are given by the equations $E(Z) = \lambda\mu$ and $var(Z) = \lambda V(\mu)$.

3.6. The Theoretical Behavior of the Prime Number Deviations

We have described the empirical behavior of the statistic $|D(x)| = |R(x) - \pi(x)$. The Tweedie convergence theorem requires for complex statistical systems with variance functions, which approximate $V(\mu) \propto \mu p$, that these systems be mathematically required to converge towards a Tweedie exponential dispersion model. The empirical demonstrations from the prime number deviations established that their variance function had this approximate behavior and the convergence requirements appeared justified.

3.7. Other Examples of Variance to Mean Power Laws and 1/f Noise

Similar variance to mean power laws have been extensively reported with nonlinear phenomena in connection with the clustering of species within their habitats (where it is known as Taylor's power law) [21], human sexual contacts in AIDS [22], pediatric leukemia cases [23] and murine experimental metastases [24]. As well, this relationship is evident with spatial and temporal data associated with long range correlations, such as temporal changes in measles incidence [25], regional blood flow heterogeneity (Bassingthwaighte's fractal scaling relation) [26], the genomic distributions of SNPs [27] and genes [28], auditory nerve spike trains [29], fluctuations in foreign exchange quotations [30], and with internet traffic [18] (where it has been called fluctuation scaling). With such diverse manifestations of this power law, the question would arise as to whether a more fundamental process could be implicated. We explain the wide manifestation of this power law by an asymptotic convergence of systems towards the Tweedie distributions [14].

The Gaussian and other stable distributions have their elementary role in probability theory consequent to the (generalized) central limit theorem, where they appear as the limiting forms for standardized sums of independent random variables with the Gaussian distribution appearing when the components have finite variances. The Tweedie convergence theorem [13] implies that a wide range of exponential dispersion models can be approximated by the Tweedie models according to a central limit-like effect. For this reason many types of non-Gaussian data will exhibit a power law relationship between the variance and the mean.

Recently, Taylor's power law and its implicit fluctuation scaling have been explained in terms of physical processes. Eisler*et al.* have provided a mean field framework that employs the summation of random variables to show that Taylor's law can be considered a consequence of a limiting theorem they relate to impact inhomogeneity [31]. Fronczak and Fronczak provided an alternative explanation that employed fluctuation dissipation in equilibrium and non-equilibrium systems to argue that Taylor's law is a consequence of the second law of thermodynamics and the action of a putative external field [32]. In the case of the prime number deviations, an explanation based on thermodynamic principles would not seem applicable.

The Tweedie convergence theorem has a more general scope than the limit theorem offered by Eisler *et al.* Indeed, the Tweedie models act as limiting distributions for a wide range of statistical models through a convergence effect that generalizes upon the central limit theorem [13,14,33,34]. Distributions within the domain of attraction of positive or extreme stable distributions can thus converge to manifest the variance to mean power law. One can hypothesize physical principles, or specific biological mechanisms, to explain the manifestation of Taylor's law, but it would seem that this mathematically demonstrable convergence property would provide a more robust and general explanation.

Sequences like the Riemann deviations $D(x)$, that express a variance to mean power law and that also obey the Tweedie PG distribution have similarly been generated from random matrices of the Gaussian unitary and orthogonal ensembles [35], the Mertens function (Equation (6)) [36], Chebyshev's function [37], as well as within the genomic distribution of SNPs along chromosome 1 of the horse [36]. Fronczak and Fronczak's derivation of the variance to mean power law, mentioned above, [32] can be shown to exactly yield the Tweedie distributions (Equations (15)–(17)) [36], and Eisler *et al.* have shown their model for impact inhomogeneity to yield equations similar to Tweedie's [31]. The Tweedie convergence effect, which has as its focus these equations, can be further extended by theorem to multiplicative statistical models which, in turn, allow for a mechanistic explanation for the genesis of multifractality [38].

$1/f$ noise is another empirical phenomenon that has been widely observed in physical and biological systems [39,40]. There is no generally accepted explanation for the appearance of $1/f$ noise, though it notably has been explained in terms of the self-organized criticality of evolutionary behavior in extended dissipative systems [41]. As noted above, $1/f$ noise is directly related to long-range correlations and, in turn, to a variance to mean power law

[18]. In this latter context, certain types of $1/f$noise can thus be viewed as consequences of the statistical convergence behavior associated with the Tweedie PG model. In fact, Taylor's law and $1/f$ noise have been demonstrated to occur together within the genomic distribution of genes and SNPs along human chromosome 1 [42].

4. CONCLUSIONS

The variance to mean power law implies a statistical self-similarity that can underlie certain fractal patterns. Related fractal patterns have been demonstrated within the distributions of the prime numbers [43] and the Riemann zeta zeros [44]. Holdom showed that the deviations $b(i)$ of the offset logarithmic integral of the ith prime number p_i, $b(i) \equiv Li(pi) - i$, exhibited scale invariant correlations that were related to the Riemann hypothesis [45]. Indeed, it is possible to express the Riemann hypothesis in terms of the behavior of a random walk on a fractal lattice [46]. In this context it should not seem surprising that Taylor's scale invariant law would manifest with the deviations $|D(x)|$.

It is the long range correlations in the distribution of the prime numbers that have been interpreted to indicate a fractal pattern [43]. The variance to mean power law is a manifestation of a fractal pattern where the power law exponent directly relates to the fractal dimension $D=H-2$ through the equation, $p=4-2D$. The assessments from Figure 2a and Figure 3, though, yielded values for p that ranged between 1 and 2. Some of this variation might reflect numerical artifact, but a component might also be attributable to the distribution of the prime numbers, itself. Indeed p can be shown to vary between 1 to 2 over different regions of the natural numbers and with different sample sizes (data not provided). When regions of a data sequence are found to have different fractal properties, as implied by the regional differences in p within the sequence of prime number deviations, that sequence would be said to be multifractal [47]. Multifractality of this nature has been described from the sequential deviations of ordered eigenvalues from the Gaussian unitary and orthogonal ensembles [38]. A current area of investigation deals with the apparent multifractality of the Riemann deviations $|D(x)|$.

The Riemann deviations $D(x)$ are intimately related to the positions of the prime numbers. Our examinations have revealed their absolute values to behave asymptotically as Tweedie probability distributions, perhaps due to a central limit-like effect, much like Erdös and Kac's demonstration of the asymptotic Gaussianity of the number of prime divisors of the integers [48]. Thus far, our evidence for a power variance to mean relationship with each of these functions is empirical, but it extends over many orders of magnitude and it adds to the growing evidence of a unifying description for disparate mathematical, physical and biological phenomena obtainable through the Tweedie exponential dispersion models. Additionally, any statistical model for sequential data designed to produce Taylor's law is mathematically required to converge towards a Tweedie model, and to yield $1/f$ noise [35].

Self-organized criticality was mentioned earlier in the introduction. This is a widely-held hypothesis used to explain the genesis of $1/f$ noise and other power law scaling behaviors [41]. Sandpile model simulations have generally been used to demonstrate $1/f$ noise and, thus, to support this hypothesis [41], yet the fluctuations evident to sandpile simulations also manifest a variance to mean power law and conform to the Tweedie compound Poisson-gamma distribution, raising the possibility that phenomena attributed to self-organized criticality can alternatively be attributed to the convergence behavior associated with the Tweedie models [15]. The irregular distribution of prime numbers thus appears to be one of many examples of Taylor's law and fluctuation scaling that are similarly implicated with this mathematical convergence behavior.

Self-organized criticality is postulated to occur within dynamical systems that naturally progress to borderline unstable and organized states, without any outside manipulation of the dynamical parameters. However, with the Riemann deviations examined here the manifestation of $1/f$ noise (the hallmark of self-organized criticality) would appear here to be possibly attributable to a mathematical convergence effect related to the central limit theorem of statistics. Further investigation into this matter could lead to a paradigm change in our understanding of processes that have conventionally been attributed to self-organized criticality.

ACKNOWLEDGMENTS

The authors gratefully acknowledge financial support from Patricia Rinaldo.

AUTHOR CONTRIBUTIONS

The authors were responsible for all aspects of this study.

CONFLICTS OF INTEREST

The authors declare no conflict of interest.

REFERENCES

1. Kotnik, T. The prime-counting function and its analytic approximations $\pi(x)$ and its approximations. *Adv. Comput. Math.* **2008**, *29*, 55–70.

2. Legendre, A.M. *Essai sur la Théorie des Nombres*, 2nd ed.; Courcier: Paris, France, 1808; p. 394. (In French)

3. Gauss, C.F. *Werke Band 2*, 1st ed.; Königliche Gesellschaft der Wissenschaften zu Göttingen: Göttingen, Germany, 1863; Volume 2, pp. 444–447. (In German)

4. Hadamard, J. Sur la distribution des zéros de la fonction et ses conséquences arithmétiques. *Bull. Soc. Math. Fr.* **1896**,*24*, 199–220. (In German).

5. De la Vallée Poussin, C.J. Recherches analytiques la théorie des nombres premiers. *Ann. Soc. Sci. Brux.* **1896**, *20*, 183–256. (In German).

6. Riemann, G.F.B. Über die anzahl der primzahlen unter einer gegebenen grösse. *Mon. Berl. Akad.* **1859**, *2*, 671–680. (In German).

7. Borwein, J.M.; Bradley, D.M.; Crandall, R.E. Computational strategies for the Riemann zeta function. *J. Comp. Appl. Math.* **2000**, *121*, 247–296.

8. Mertens, F. Über eine zahlentheoretische funktion. *Akad. Wiss. Math. Nat. Kleine Sitz.* **1897**, *106*, 761–830. (In German).

9. Titchmarsh, E.C. *The Theory of the Riemann Zeta-Function*, 2nd ed.; The Clarendon Oxford University Press: New York, NY, USA, 1986.

10. Gamba, Z.; Hernando, A.; Romanelli, L. Are prime numbers regularly ordered? *Phys. Lett. A* **1990**, *145*, 106–108.

11. Wolf, M. 1/f noise in the distribution of prime numbers. *Phys. A* **1997**, *241*, 493–499.

12. Kriecherbauer, T.; Marklof, J.; Soshnikov, A. Random matrices and quantum chaos. *Proc. Natl. Acad. Sci. USA* **2001**, *98*, 10531–10532.

13. Jørgensen, B. *The Theory of Dispersion Models*; Chapman & Hall: London, UK, 1997.

14. Jørgensen, B.; Martínez, J.R.; Tsao, M. Asymptotic behaviour of the variance function. *Scand. J. Stat.* **1994**, *213*, 223–243.

15. Kendal, W.S. Self-organized criticality attributed to a central limit-like convergence effect. *Phys. A* **2015**, *412*, 141–150.

16. Riesel, H.; Göhl, G. Some calculations related to Riemann's prime number formula. *Math. Comput.* **1970**, *24*, 969–983.

17. Leland, W.E.; Taqqu, M.S.; Willinger, W.; Wilson, D.V. On the self-similar nature of ethernet traffic. *IEE/ACM Trans. Netw.* **1994**, *2*, 1–15.

18. Tsybakov, B.; Georganas, N.D. On self-similar traffic in atm queues: Definitions, overflow probability bound, and cell delay distribution. *IEEE/ACM Trans. Netw.* **1997**, *5*, 397–409.

19. McQuarrie, D.A. *Statistical Mechanics*; Harper & Row: New York, NY, USA, 1976; pp. 553–561.

20. Tweedie, M.C.K. An index which distinguishes between some important exponential families. In *Statistics: Applications and New Directions*, Proceedings of the Indian Statistical Institute Golden Jubilee International Conference, Calcutta, India, 27

September–1 October 1982; Ghosh, J.K., Roy, J., Eds.; Indian Statistical Institute: Calcutta, India, 1984; pp. 579–604.

21. Taylor, L.R. Aggregation, variance and the mean. *Nature* **1961**, *189*, 732–735.

22. Anderson, R.M.; May, R.M. Epidemiological parameters of HIV transmission. *Nature* **1988**, *333*, 514–519.

23. Philippe, P. The scale-invariant spatial clustering of leukemia in San Francisco. *J. Theor. Biol.* **1999**, *199*, 371–381.

24. Kendal, W.S.; Frost, P. Experimental metastasis: A novel application of the variance-to-mean power function. *J. Natl. Cancer Inst.* **1987**, *79*, 1113–1115.

25. Rhodes, C.J.; Anderson, R.M. A scaling analysis of measles epidemics in a small population. *Philos. Trans. R. Soc. Lond. B Biol. Sci.* **1996**, *351*, 1679–1688.

26. Kendal, W.S. A stochastic model for the self-similar heterogeneity of regional organ blood flow. *Proc. Natl. Acad. Sci. USA* **2001**, *98*, 837–841.

27. Kendal, W.S. An exponential dispersion model for the distribution of human single nucleotide polymorphisms. *Mol. Biol. Evol.* **2003**, *20*, 579–590.

28. Kendal, W.S. A scale invariant clustering of genes on human chromosome 7. *BMC Evol. Biol.* **2004**, *4*, 3.

29. Lowen, S.B.; Teich, M.C. The periodogram and Allan variance reveal fractal exponents greater than unitiy in auditory-nerve spike trains. *J. Acoust. Soc. Am.* **1996**, *99*, 3585–3591.

30. Sato, A.H.; Nishimura, M.; Holyst, J.A. Fluctuation scaling of quotation activities in the foreign exchange market.*Phys. A* **2010**, *389*, 2793–2804.

31. Eisler, Z.; Bartos, I.; Kertesz, J. Fluctuation scaling in complex systems: Taylor's law and beyond. *Adv. Phys.* **2008**, *57*, 89–142.

32. Fronczak, A.; Fronczak, P. Origins of Taylor's power law for fluctuation scaling in complex systems. *Phys. Rev. E* **2010**,*81*, 066112.

33. Jørgensen, B.; Vinogradov, V. Convergence to tweedie models and related topics. In *Advances on Theoretical and Methodological Aspects of Probability and Statistics*; Balakrishnan, N., Ed.; Taylor & Francis: New York, NY, USA, 2002; pp. 473–489.

34. Vinogradov, V.; Jørgensen, B.; Wentzell, B.D. From extreme stable laws to Tweedie exponential dispersion models. In*Stochastic Models*, Proceedings of the International Conference on Stochastic Models, Ottawa, ON, Canada, 10–13 June 1998; Dawson, D.A., Gorostiva, L., Ivanoff, G., Eds.; American Mathematical Society: Providence, RI, USA, 1999; pp. 435–443.

35. Kendal, W.S.; Jørgensen, B. Tweedie convergence: A mathematical basis for Taylor's power law, $1/f$ noise and multifractality. *Phys. Rev. E* **2011**, *84*, 066120.

36. Kendal, W.S.; Jørgensen, B. Taylor's power law and fluctuation scaling explained by a central-limit-like convergence.*Phys. Rev. E* **2011**, *83*, 066115.

37. Kendal, W.S. Fluctuation scaling and $1/f$ noise: Shared origins from the Tweedie family of statistical distributions. *J. Basic Appl. Phys.* **2013**, *2*, 40–49.

38. Kendal, W.S. Multifractality attributed to dual central limit-like convergence effects. *Phys. A* **2014**, *401*, 22–33.

39. Dutta, P.; Horn, P.M. Low-frequency fluctuations in solids: 1/*f* noise. *Rev. Mod. Phys.* **1981**, *53*, 497–516.

40. Musha, T. 1/f Fluctuations in Biological Systems. In Proceedings of the Sixth International Conference on Noise in Physical Systems, Washington, DC, USA, 6–10 April 1981; pp. 143–146.

41. Bak, P.; Tang, C.; Wiesenfeld, K. Self-organized criticality: An explanation of the 1/*f* noise. *Phys. Rev. Lett.* **1987**, *59*, 381–384.

42. Kendal, W.S. Scale invariant correlations between genes and SNPs on human chromosome 1 reveal potential evolutionary mechanisms. *J. Theor. Biol.* **2007**, *245*, 329–340.

43. Cattani, C. Fractal patterns in prime numbers distribution. *Lect. Notes Comput. Sci.* **2010**, *6017*, 164–176.

44. Van Zyl, B.P.; Hutchinson, D.A.W. Riemann zeros, prime numbers, and fractal potentials. *Phys. Rev. E* **2003**, *67*, 066211.

45. Holdom, B. Scale-invariant correlations and the distribution of prime numbers. *J. Phys. A Math. Theor.* **2009**, *42*, 345102.

46. Shlesinger, M.F. On the Riemann hypothesis: A fractal random walk approach. *Phys. A* **1986**, *138*, 310–319.

47. Stanley, H.E.; Meakin, P. Multifractal phenomena in physics and chemistry. *Nature* **1988**, *335*, 405–409.

48. Erdös, P.; Kac, M. The Gaussian law of errors in the theory of additive number theoretic functions. *Am. J. Math.* **1940**,*62*, 738–742.

CHAPTER 9

Generalized q-Stirling Numbers and Their Interpolation Functions

Hacer Ozden [1,*], Ismail Naci Cangul [1] and Yilmaz Simsek [2]

[1]Faculty of Arts and Science, Department of Mathematics, Uludag University, TR-16059, Bursa, Turkey

[2]Faculty of Science, Department of Mathematics, University of Akdeniz, TR-07058, Antalya, Turkey

ABSTRACT

In this paper, we define the generating functions for the generalized q-Stirling numbers of the second kind. By applying Mellin transform to these functions, we construct interpolation functions of these numbers at negative integers. We also derive some identities and relations related to q-Bernoulli numbers and polynomials and q-Stirling numbers of the second kind.

KEYWORDS

q-Bernoulli numbers and polynomials; generalized q-Stirling numbers of the second kind; q-zeta function

1. INTRODUCTION, DEFINITIONS AND NOTATIONS

q-Stirling numbers of the second kind were first defined by Carlitz [1]. After Carlitz's paper, many combinatorial papers have centered around the q-analogue, the earliest by Milne [2]; (among others) also see [3,4,5,6,7,8,9,10,11,12,13,14,15,16,17,18,19,20].

In [16], Simsek studied the generating functions of the fermionic and deformic Stirling numbers. By applying the derivative operator dndtn||t=0 to these functions, he constructed interpolation functions of these numbers at negative integers.

It is well known that the Stirling numbers of the second kind S(n,k) are defined by means of the generating function [16,17,18,19,20,21]:

$$F_S(t,k) = \frac{(-1)^k}{k!}(1-e^t)^k = \sum_{n=0}^{\infty} S(n,k)\frac{t^n}{n!} \tag{1}$$

It is also well known that the usual Stirling numbers of the second kind S(α)(n,k) are defined by means of the generating function [16,17,18,19,20,21]:

$$F_\alpha(t,k) = \frac{(-1)^k}{k!} e^{\alpha t}(1-e^t)^k = \sum_{n=0}^{\infty} S^{(\alpha)}(n,k)\frac{t^n}{n!} \qquad (2)$$

And

$$F_\alpha^r(t,k) = \frac{(-1)^k}{k!} e^{\alpha t}(1-e^{rt})^k = \sum_{n=0}^{\infty} S^{(\frac{\alpha}{r})}(n,k)\frac{t^n}{n!} \qquad (3)$$

Let q∈C with |q|<1. Some well known results related to the *q*-integers are given by (see for detail [1,2,3,4,5,6,7,8,9,10,11, 12,13,14,15,16,17,18,19,20,21,22,23,24,25,26,27,28]):

$$\begin{aligned} [n,q] &= [n] = \frac{1-q^n}{1-q} \\ [k+j] &= [k] + q^k[j] \\ [kj] &= [k][j,q^k] \end{aligned}$$

And

$$[n]! = [n][n-1]...[2][1],\ [0]! = 1 \text{ and } \binom{n}{k}_q = \frac{[n]!}{[n-k]![k]!}$$

Note that limq→1[n]=n, 1,2,3,4,5,6,7,8,9,10,11,12, 13,14,15,16,17, 18,19,20,21,22,23,24,25,26,27,28].

Generating functions of the *q*-Stirling numbers of the second kind were defined in [8]:

$$F_q(t) = q^{\binom{k}{2}} \frac{1}{[k]!} \sum_{j=0}^{k} \binom{k}{j} (-1)^{k-j} q^{\binom{k-j}{2}} \exp([j]t) \qquad (4)$$

And

$$[x]^n = \sum_{k=0}^{n} \left(\begin{array}{c} m \\ n \end{array} \right)_q [k]!q^{\left(\begin{array}{c} k \\ 2 \end{array} \right)} S_2(n,k,q)$$

By the above equation, we have [1,8]

$$S_2(n,k,q) = q^{-\left(\begin{array}{c} k \\ 2 \end{array} \right)} \frac{1}{[k]!} \sum_{j=0}^{k} (-1)^j q^{\left(\begin{array}{c} j \\ 2 \end{array} \right)} \left(\begin{array}{c} k \\ j \end{array} \right)_q [k-j]^n$$

2. NEW GENERATING FUNCTIONS FOR *Q*-STIRLING NUMBERS OF THE SECOND KIND

Here, by using the same method of Simsek [16,18], we construct interpolation functions for the generalized q-Stirling numbers of the second kind. We shall define new functions to interpolate the second kind q-Stirling numbers. We define q-version of Equation (1) and Equation (2) functions. Generalized q-Stirling numbers of the second kind are defined by means of the following generating functions:

$$\begin{aligned} F_{q,\alpha}(t,k) &= \frac{1}{[k]!} \left(-\sum_{n=0}^{\infty} q^n \exp\left([n - \frac{\alpha}{k}]t \right) \right)^{-k} \\ &= \sum_{n=0}^{\infty} S^{(\alpha)}(n,k,q) \frac{t^n}{n!} \end{aligned} \tag{5}$$

Or

$$F^*_{q,\alpha}(t,k) = \frac{1}{[k]!}\sum_{j=0}^{k}(-1)^{k+j}\binom{k}{j}q^k e^{[\alpha+j]t}$$
$$= \sum_{n=0}^{\infty}\frac{1}{[k]!}\sum_{j=0}^{k}(-1)^{k+j}\binom{k}{j}q^k\frac{[\alpha+j]^n t^n}{n!} \tag{6}$$

And

$$F^*_{q,\alpha}(t) = \sum_{n=0}^{\infty} S^{(\alpha)}(n,k,q)\frac{t^n}{n!}$$

By comparing the coefficients of tnn! on both sides of the above equations, we easily obtain that

$$S^{(\alpha)}(n,k,q) = \frac{1}{[k]!}\sum_{j=0}^{k}(-1)^{k+j}\binom{k}{j}q^k[\alpha+j]^n$$
$$= \frac{(1-q)^{-n}}{[k]!}\sum_{j=0}^{k}\sum_{c=0}^{n}(-1)^{k+j+c}\binom{k}{j}\binom{n}{c}q^{k+(\alpha+j)c}$$

Observe that when q→1, Equations (5) and (6) reduce to Equation (2). When q→1 in Equation (5), we have

$$\lim_{q\to 1} F_{q,\alpha}(t,k) = \frac{1}{k!}\left(-\sum_{n=0}^{\infty}q^n \exp\left(nt-\frac{\alpha}{k}t\right)\right)^{-k}$$
$$\sum_{n=0}^{\infty} S^{(\alpha)}(n,k)\frac{t^n}{n!} = \frac{(-1)^k}{k!}\left(e^t-1\right)^k = \frac{(-1)^k\exp(\alpha t)}{k!}\sum_{y=0}^{k}\binom{k}{y}(-1)^y\exp(yt)$$
$$= \sum_{n=0}^{\infty}\left(\frac{(-1)^k\exp(\alpha t)}{k!}\sum_{y=0}^{k}\binom{k}{y}(-1)^y y^n\right)\frac{t^n}{n!}$$

Here we use the binomial expansion and the fact that

$$e^x = \sum_{n=0}^{\infty} \frac{x^n}{n!}$$

By comparing the coefficients of $\frac{t^n}{n!}$ on both sides of the above equations, we easily obtain that

$$S(n,k) = \frac{(-1)^k \exp(\alpha t)}{k!} \sum_{y=0}^{k} \binom{k}{y} (-1)^y y^n$$

We also see that

$$S^{(\alpha)}(n,k) = \frac{(-1)^k}{k!} \sum_{y=0}^{k} \binom{k}{y} (-1)^y (\alpha+y)^n$$

with recurrence relation in [21]

$$S^{(\alpha)}(n,k) = S^{(\alpha)}(n-1,k-1) + (k+\alpha) S^{(\alpha)}(n-1,k)$$

We also define the following generating function which is generalized Equation (6):

$$\begin{aligned} F^*_{q,\frac{\alpha}{r}}(t,k) &= \frac{1}{[k]!} \sum_{j=0}^{k} (-1)^{k+j} \binom{k}{j} q^k e^{[\alpha+rj]t} \\ &= \sum_{n=0}^{\infty} S^{(\frac{\alpha}{r})}(n,k,q) \frac{t^n}{n!} \end{aligned} \tag{7}$$

By Equation (7), we obtain

$$S^{(\frac{\alpha}{r})}(n,k,q)=\frac{1}{[k]!}\sum_{j=0}^{k}\sum_{d=0}^{n}(-1)^{k+j}\begin{pmatrix}k\\j\end{pmatrix}\begin{pmatrix}n\\d\end{pmatrix}q^{k+d\alpha}[\alpha]^{n-d}[rj]^{d}$$

By using Pb. 189 in [24], we can write

$$S(n,k)=\frac{1}{k!}\left(k^{n}-\begin{pmatrix}k\\1\end{pmatrix}(k-1)^{n}+\begin{pmatrix}k\\2\end{pmatrix}(k-2)^{n}-\ldots+(-1)^{k}0^{n}\right),n\geq 1$$

(8)

We give the q-version of the above equation as follows

$$S(n,k,q)=\frac{1}{[k]!}\left([k]^{n}-\begin{pmatrix}k\\1\end{pmatrix}_{q}[k-1]^{n}+\begin{pmatrix}k\\2\end{pmatrix}_{q}[k-2]^{n}\ldots+[-1]^{k}[0]^{n}\right)$$

3. SOME SPECIAL ZETA FUNCTIONS

Throughout this section, let s∈C with Res>1. By using the same method of Simsek [16,18], we construct interpolation functions for the generalized q-Stirling numbers of the second kind. By applying the Mellin transform to Equation (4), we have

$$\frac{1}{\Gamma(s)}\int_{0}^{\infty}t^{s-1}F_{q}(-t)dt$$

$$=q^{\begin{pmatrix}k\\2\end{pmatrix}}\frac{1}{[k]!}\sum_{j=0}^{k}\begin{pmatrix}k\\j\end{pmatrix}(-1)^{k-j}q^{\begin{pmatrix}k-j\\2\end{pmatrix}}\frac{1}{\Gamma(s)}\int_{0}^{\infty}t^{s-1}\exp(-[j]t)dt$$

$$=q^{\begin{pmatrix}k\\2\end{pmatrix}}\frac{1}{[k]!}\sum_{j=0}^{k}\begin{pmatrix}k\\j\end{pmatrix}(-1)^{k-j}q^{\begin{pmatrix}k-j\\2\end{pmatrix}}\frac{1}{[j]^{s-1}}$$

Thus we define the following zeta function:

Definition 1 *Let* s∈C *and* k∈Z+, *the set of positive integers.*

$$z(s,k) = q^{\binom{k}{2}} \frac{1}{[k]!} \sum_{j=0}^{k} \binom{k}{j} (-1)^{k-j} q^{\binom{k-j}{2}} \frac{1}{[j]^{s-1}}$$

By substituting s=1−n into above definition, we have

$$z_q(1-n,k) = q^{\binom{k}{2}} \frac{1}{[k]!} \sum_{j=0}^{k} \binom{k}{j} (-1)^{k-j} q^{\binom{k-j}{2}} [j]^n$$

Using the above relation, we arrive at the following result:

Theorem 1

Let n *and* k *be positive integers. Then*

$$z_q(1-n,k) = S(n,k,q)$$

By applying the Mellin transform to the Equation (2), we have

$$\frac{1}{\Gamma(s)} \int_0^\infty t^{s-1} F_\alpha(-t,k)\,dt = \frac{(-1)^k}{k!} \sum_{j=0}^{k} \binom{k}{j} \frac{(-1)^j}{(j+\alpha)^{s-1}}$$

So we have the following definition:

Definition 2

Let s∈C *and* k∈Z+.

$$z^{\alpha}(s,k)=\frac{(-1)^k}{k!}\sum_{j=0}^{k}\begin{pmatrix}k\\j\end{pmatrix}\frac{(-1)^j}{(j+\alpha)^{s-1}}$$

Remark 1 *If* α=0 *above, then we have*

$$z^{0}(s,k)=\frac{(-1)^k}{k!}\sum_{j=0}^{k}\begin{pmatrix}k\\j\end{pmatrix}\frac{(-1)^j}{j^{s-1}}$$

For s=1−n, n∈Z+ above equation, we obtain

$$\begin{aligned}z^{\alpha}(1-n,k) &= \frac{(-1)^k}{k!}\sum_{j=0}^{k}\begin{pmatrix}k\\j\end{pmatrix}\frac{(-1)^j}{(j+\alpha)^{-n}}\\ &= S^{(\alpha)}(n,k)\end{aligned}$$

Therefore, we arrive at the following result:

Corollary 1

Let n∈Z+. *Then we have*

$$z^{\alpha}(1-n,k)=S^{(\alpha)}(n,k)$$

Remark 2

$$\begin{aligned}(e^t-1)^k &= \sum_{j=0}^{k}\begin{pmatrix}k\\j\end{pmatrix}(-1)^j e^{jt}\\ &= \sum_{n=0}^{\infty}\left(k^n\begin{pmatrix}k\\0\end{pmatrix}-\begin{pmatrix}k\\j\end{pmatrix}(k-1)^n+\ldots+\begin{pmatrix}k\\k\end{pmatrix}(-1)^k(k-k)^n\right)\frac{t^n}{n!}\end{aligned}$$

By using Equation (8), we have

$$(e^t - 1)^k = k! \sum_{n=0}^{\infty} z^0(1-n,k)\frac{t^n}{n!}$$

By applying the Mellin transform to Equations (6) and (7), we define the following functions, respectively:

$$\frac{1}{\Gamma(s)}\int_0^{\infty} t^{s-1}F^*_{q,\alpha}(-t,k)dt = z_q^{\alpha}(s,k)$$
$$\frac{1}{\Gamma(s)}\int_0^{\infty} t^{s-1}F^*_{q,\frac{\alpha}{r}}(-t,k)dt = z_q^{\frac{\alpha}{r}}(s,k)$$

The above functions interpolate the numbers $S^{(\alpha)}(n,k,q)$ and $S^{(\frac{\alpha}{r})}(n,k,q)$ at negative integers, respectively.

4. RELATIONS BETWEEN BERNOULLI NUMBERS OF ORDER *K* AND STIRLING NUMBERS OF THE SECOND KIND

Let

$$F_B(t) = \frac{t^k}{(e^t-1)^k} = \sum_{n=0}^{\infty} B_n^{(k)}\frac{t^n}{n!} \tag{9}$$

where the coefficients B(k)n are called Bernoulli numbers of order *k* [19,20,28]. By Equation (1), we have

$$F_S(t,k) = \frac{1}{k!}(e^t - 1)^k \tag{10}$$

By using Equations (9) and (10), relation between FB(t) and FS(t) is given by

$$F_B(t)F_S(t) = \frac{t^k}{k!}$$

By using the above relation, we have

$$\sum_{n=0}^{\infty} B_n^{(k)} \frac{t^n}{n!} \sum_{n=0}^{\infty} S(n,k)\frac{t^n}{n!} = \frac{t^k}{k!}$$

By using Cauchy product above, we get

$$\sum_{n=0}^{\infty}\sum_{j=0}^{n} B_j^{(k)} S(n-j,k)\frac{1}{k!}\frac{t^n}{(n-k)!} = \frac{t^k}{k!}$$

By comparing coefficients of tk in both sides of the above equation, we arrive at the following theorem:

Theorem 2

Let n,k∈N. *We have*

$$n!k!\sum_{j=0}^{n}\binom{n}{j} B_j^{(k)} S(n-j,k) = \begin{cases} 1, \text{ if } n = k \\ 0, \text{ if } n \neq k \end{cases}$$

Remark 3

The Barnes' type multiple Changhee q-*Bernoulli polynomials are defined by means of the following generating function (see for details [28]):*

$$\begin{aligned} G_q^{(k)}(w,t \mid w_1, w_2, \ldots, w_k) &= (-t)^k \left(\prod_{i=1}^{k} w_i\right) \sum_{n_1, n_2, \ldots, n_k=0}^{\infty} q^{w+n_1w_1+n_2w_2+\ldots+n_kw_k} e^{[w+n_1w_1+n_2w_2+\ldots+n_kw_k]t} \\ &= \sum_{n=0}^{\infty} \frac{B_n^{(k)}(w:q \mid w_1, w_2, \ldots, w_k)t^n}{n!} \quad (\mid t \mid < 2\pi) \end{aligned} \tag{11}$$

with as usual,

$$\sum_{n_1, n_2, \ldots, n_k=0}^{\infty} = \sum_{n_1=0}^{\infty} \sum_{n_2=0}^{\infty} \cdots \sum_{n_k=0}^{\infty}$$

It follows from Equation (11) that

$$\lim_{q \to 1} G_q^{(k)}(w,t \mid w_1, w_2, \ldots, w_k) = \frac{e^{tw}(tw_1)(tw_2)\ldots(tw_k)}{(e^{tw_1}-1)(e^{tw_2}-1)\ldots(e^{tw_k}-1)} \tag{12}$$

This gives the generating function of Barnes' type multiple Bernoulli numbers. Thus we get the following limit relationship:

$$\lim_{q \to 1} B_n^{(k)}(w:q \mid w_1, w_2, \ldots, w_k) = B_n^{(k)}(w \mid w_1, w_2, \ldots, w_k)$$

This gives the Barnes' type multiple Bernoulli numbers as a limit when q *approaches 1.*

$$\lim_{q \to 1} G_q^{(k)}(0,t \mid 1,1,\ldots,1) = \frac{t^k}{(e^t-1)^k} = \sum_{n=0}^{\infty} B_n^{(k)} \frac{t^n}{n!}$$

Using Equation (12), we define

$$F_B^{(n)}(t \mid w_1, w_2, ..., w_k) = \frac{t^{kn}(w_1 w_2 ... w_k)^n}{(e^{tw_1} - 1)^n (e^{tw_2} - 1)^n ... (e^{tw_k} - 1)^n}$$
$$= \sum_{n=0}^{\infty} B_n^{(k,n)}(0 \mid w_1, w_2, ..., w_k) \frac{t^n}{n!}$$

(13)

Observe that when $n = 1$, $B_n^{(k,n)}(0 \mid w_1, w_2, ..., w_k)$ reduces to $B_n^{(k)}(0 \mid w_1, w_2, ..., w_k)$. We define

$$F_S^{(n)}(t \mid w_1, w_2, ..., w_k) = \left(\frac{1}{k!}\right)^n (e^{tw_1} - 1)^n (e^{tw_2} - 1)^n ... (e^{tw_k} - 1)^n$$
$$= \sum_{n=0}^{\infty} Y^{(n)}(n, k \mid w_1, w_2, ..., w_k) \frac{t^n}{n!}$$

(14)

By Equations (13) and (14), we have

$$(\frac{k!}{w_1 w_2 ... w_k})^n F_S^{(n)}(t \mid w_1, w_2, ..., w_k) F_B^{(n)}(t \mid w_1, w_2, ..., w_k) = t^{kn}$$

By using the above equation, we have

$$(\frac{k!}{w_1 w_2 ... w_k})^n \sum_{m=0}^{\infty} Y^{(n)}(m, k \mid w_1, w_2, ..., w_k)$$
$$\frac{t^m}{m!} \sum_{m=0}^{\infty} B_m^{(k,n)}(0 \mid w_1, w_2, ..., w_k) \frac{t^m}{m!} = t^{kn}$$

By applying the Cauchy product to the above, we arrive at the following theorem, which is the generalized form of Theorem 2:

Theorem 3

Let n,k∈N. *We have*

$$m!(\frac{k!}{w_1w_2...w_k})^n\sum_{j=0}^{m}\begin{pmatrix} m \\ j \end{pmatrix}B_n^{(k,n)}(0 \quad | \quad w_1,w_2,...,w_k)Y^{(n)}(m-j,k \mid w_1,w_2,...,w_k)$$
$$= \begin{cases} 1, \mathit{if}\, m=nk \\ 0, \mathit{if}\, m\neq nk \end{cases}$$

Observe that

$$F_S^{(1)}(t \quad | \quad 1,1,...,1) = \sum_{n=0}^{\infty}Y^{(1)}(n,k \mid 1,1,...,1)\frac{t^n}{n!}$$
$$= \frac{1}{k!}(e^t-1)^k$$
$$= \sum_{n=0}^{\infty}S(n,k)\frac{t^n}{n!}$$

Thus we have

$$Y^{(1)}(n,k \mid 1,1,...,1) = S(n,k)$$

5. CONCLUSIONS

q-Stirling numbers of the second kind arise in many different generating functions for various statistical partitions. The theory of *q*-Stirling numbers is enriched by combinatorial interpretations. By using these numbers, one can investigate orthogonality relations, recurrences, explicit expressions, and generating functions for the generalized (q-) Stirling numbers. Recently, many authors have generalized the Stirling numbers by differential operators. The Stirling numbers are related to Newton's interpolation, (q-) Lah numbers, exponential generating functions, q-calculus and related topics, combinatorial enumeration problems, Binomial coefficients and Bell numbers.

ACKNOWLEDGEMENTS

This paper was supported by the Scientific Research Project Administration of Akdeniz University and by the Scientific Research Fund of Uludag University, Project numbers F-2012/16 and F-2012/19. We would like to thank to referees for their valuable comments and many references to suggest for adding this present papers.

REFERENCES

1. Carlitz, L. q-Bernoulli numbers and polynomials. *Duke Math.* **1948**, *15*, 987–1000.
2. Milne, S.C. A q-analog of restricted growth functions, Dobinski's equality, and Charlier polynomials. *Trans. Amer. Math. Soc.* **1978**, *245*, 89–118.
3. Ehrenborg, R. Determinants involving q-Stirling numbers. *Advan. Appl. Math.* **2003**, *31*, 630–642.
4. Ernst, T. *A Comprehensive Treatment of q-Calculus*; Birkhauser/ Springer: Basel, Switzerland, 2012.

5. Gould, H.W. The q-Stirling numbers of first and second kinds. *Duke Math. J.* **1961**, *28*, 281–289.

6. Hsu, L.C.; Shiue, P.J.-S. A unified approach to generalized Stirling numbers. *Adv. Appl. Math.* **1998**, *20*, 366–384.

7. Katriel, J. Stirling numbers identity: Interconsistency of q-analogues. *J. Phys.* **1998**, *31*, 3559–3672.

8. Kim, T. q-Volkenborn integration. *Russ. J. Math. Phys.* **2002**, *19*, 288–299.

9. Kim, T. q-Bernoulli numbers and polynomials associated with Gaussian binomial coefficients. *Russ. J. Math. Phys.***2008**, *15*, 51–57.

10. Kwasniewski, A.K. q-Poisson, q-Dobinski, q-Rota and q-coherent states-a second fortieth anniversary memoir. *Proc. Jangjeon Math. Soc.* **2004**, *7*, 95–98.

11. Kwasniewski, A.K. On psi-umbral extension of Stirling numbers and Dobinski-like formulas. *Advan. Stud. Contemp. Math.* **2006**, *12*, 73–100.

12. Mansour, T.; Schork, M.; Shattuck, M. On a new family of generalized Stirling and Bell numbers. *Electron. J. Comb.***2011**, *18*, 1–33.

13. Mansour, T.; Schork, M.; Shattuck, M. The generalized Stirling and Bell numbers revisited. *J. Integer Seq.* **2012**, *15*, 1–47.

14. Mansour, T.; Schork, M.; Shattuck, M. On the Stirling numbers associated with the meromorphic Weyl algebra. *Appl. Math. Lett.* **2012**, *25*, 1767–1771.

15. Ryoo, C.S.; Kim, T. A new identities on the q-Bernoulli numbers and polynomials. *Adv. Stud. Contemp. Math.* **2011**, *21*, 161–169.

16. Simsek, Y. On q-Deformed Stirling numbers. *Int. J. Math. Comput.* **2012**, *15*, 70–80.

17. Simsek, Y. Generating functions for generalized Stirling type numbers, Array type polynomials, Eulerian type polynomials and their application. 2011; arxiv:1111.3848v2. Available online: http://arxiv.org/abs/1111.3848 (accessed on 5 February 2013).

18. Simsek, Y. *Interpolation Function of Generalized q-Bernstein Type Polynomials and Their Application, Curve and Surface*; Springer Verlag: Berlin/Heidelberg, Germany, 2011; pp. 647–662.

19. Srivastava, H.M. Some generalizations and basic (or q-) extensions of the Bernoulli, Euler and Genocchi polynomials. *Appl. Math. Inform. Sci.* **2011**, *5*, 390–444.

20. Srivastava, H.M.; Choi, J. *Zeta and q-Zeta Functions and Associated Series and Integrals*; Elsevier Science Publishers: Amsterdam, The Netherlands; London, UK; New York, NY, USA, 2012.

21. Cakic, M.P.; Milovanovic, G.V. On generalized Stirling numbers and polynomials. *Math. Balk.* **2004**, *18*, 241–248.

22. Choi, J.; Kim, T.; Kim, Y.H. A note on the extended q-Bernoulli numbers and polynomials. *Adv. Stud. Contemp. Math. (Kyungshang)* **2011**, *21*, 351–354.

23. Kim, T. Non-Archimedean q-integrals associated with multiple Changhee q-Bernoulli polynomials. *Russ. J. Math. Phys.* **2003**, *10*, 91–98.

24. Polya, G.; Szego, G. *Problems and Theorems in Analysis I*; Springer Verlag: Berlin/Heidelberg, Germany, 1973.

25. Rassias, T.M.; Srivastava, H.M. Some classes of infinite series associated with the Riemann zeta and polygamma functions and

generalized harmonic numbers. *Appl. Math. Comput.* **2002**, *131*, 593–605.

26. Simsek, Y. q-Analogue of the twisted l-series and q-twisted Euler numbers. *J. Number Theory* **2005**, *110*, 267–278.

27. Simsek, Y. Twisted (h,q)-Bernoulli numbers and polynomials related to twisted (h,q)-zeta function and L-function. *J. Math. Anal. Appl.* **2006**, *324*, 790–804.

28. Srivastava, H.M.; Kim, T.; Simsek, Y. q-Bernoulli numbers and polynomials associated with multiple q-zeta functions and basic L-series. *Russ. J. Math. Phys.* **2005**, *12*, 241–268.

CHAPTER 10

On θ-Congruent Numbers, Rational Squares in Arithmetic Progressions, Concordant Forms and Elliptic Curves

Erich Selder [1] and Karlheinz Spindler [2,*]

[1]Fachbereich 2, Fachhochschule Frankfurt, Nibelungenplatz 1, D-60318 Frankfurt am Main, Germany

[2]Fachbereich Architektur und Bauingenieurwesen, Studiengang Angewandte Mathematik, Hochschule RheinMain, Kurt-Schumacher-Ring 18, D-65197 Wiesbaden, Germany

ABSTRACT

The correspondence between right triangles with rational sides, triplets of rational squares in arithmetic succession and integral solutions of certain quadratic forms is well-known. We show how this correspondence can be extended to the generalized notions of rational θ-triangles, rational squares occurring in arithmetic progressions and concordant forms. In our approach we establish one-to-one mappings to rational points on certain elliptic curves and examine in detail the role of solutions of the θ-congruent number problem and the concordant form problem associated with nontrivial torsion points on the corresponding elliptic curves. This approach allows us to combine and extend some disjoint results obtained by a number of authors, to clarify some statements in the literature and to answer some hitherto open questions.

KEYWORDS

elliptic curves; concordant forms; θ-congruent numbers

1. INTRODUCTION

The following definition dates back to Euler ([1]; see also [2]).

Definition 1. *Two quadratic forms $X^2 + mY^2$ and $X^2 + nY^2$ (where $m, n \in \mathbb{Z} \setminus \{0\}$ with $m \neq n$) are called concordant if the system*

$$X^2 + mY^2 = Z^2, \quad X^2 + nY^2 = W^2$$

admits a nontrivial solution $(X, Y, Z, W) \in \mathbb{Z}^4$, where nontriviality means that $Y \neq 0$. This is equivalent to saying that there are solutions $(X, Y, Z, W) \in \mathbb{P}_3(\mathbb{Q})$ other than $(1, 0, \pm1, \pm1)$. Thus (X, Y, Z, W) is a trivial solution if and only if it is a solution for any pair (m, n).

While this definition essentially coincides with the one mentioned in the Introduction of [3] (in which $XY \neq 0$ is required), we note that in the sequel of [3] and also in the subsequent paper [4] only solutions in which all components are nonzero are studied. We emphasize that we consider not only such solutions to be nontrivial, but all solutions other than those of the form $(\pm X, 0, \pm X, \pm X)$. (Note that a system with a solution (X, Y, Z, W) for which $X = 0$ can be easily transformed into a system with a solution (X, Y, Z, W) for which $Z = 0$ or $W = 0$ but $X \neq 0$; hence allowing $Z = 0$ and $W = 0$ essentially also allows $X = 0$.)

It is easily verified that every system of concordant forms is equivalent to one in which the coefficients m and n have different signs; hence we may always assume that $m < 0$ and $n > 0$. After factoring out the greatest common divisor of the coefficients, this leads us to consider quadratic forms $X^2 - pkY^2$ and $X^2 + qkY^2$ where $k, p, q \in \mathbb{N}$ with p and q coprime. Concordant forms in this form tie up nicely with rational squares occurring in arithmetic progressions. In fact, if $\alpha^2 < \beta^2 < \gamma^2$ are squares of rational numbers that occur in an arithmetic progression of (maximally chosen) step size k, then there are coprime numbers $p, q \in \mathbb{N}$ such that $\alpha^2 = \beta^2 - pk$ and $\gamma^2 = \beta^2 + qk$, which means exactly that the forms $X^2 - pkY^2$ and $X^2 + qkY^2$ are concordant. To have a precise terminology available, let us give a formal definition.

Definition 2. *A triplet $(p, q, k) \in \mathbb{N}\times\mathbb{N}\times\mathbb{N}$ where p, q are coprime is called a solution of the concordant form problem if and only if the quadratic forms $X^2 - pkY^2$ and $X^2 + qkY^2$ are concordant; i.e., if and only if there is an arithmetic progression of (maximal) step size k containing three rational squares, where the lowest (respectively the highest) is separated from the intermediate one by p times (respectively q times) the step size.*

Obviously, a triplet (p, q, k) is a solution of the concordant form problem if and only if (p, q, a^2k) is for any $a \in \mathbb{N}$ (since the factor a can be subsumed into Y); hence it is sufficient to study solutions (p, q, k) where k is squarefree. We note that arithmetic progressions of squares have been studied not just over the rationals, but over number fields (*cf.* [5–8]). While in these approaches the goal was to find (maximal) uninterrupted arithmetic progressions of squares in the given base field, we focus on rational squares that occur in arithmetic progressions, but not necessarily in immediate succession. We now relate the concordant form problem to a different problem which is cast in geometric rather than arithmetic language. The concept of congruent numbers (see [9,10]) has been extended to that of a t-congruent number (*cf.* [11]) and even more generally to that of a θ-congruent number (*cf.* [12–16]); for an overview see [11]. Even though the concept of a t-congruent number is more natural from a geometric point of view (arising from the search for triangles with rational sides and rational area), the more general concept of a θ-congruent number is more relevant for the purposes of this paper.

Definition 3. *Given an angle $\theta \in (0, \pi)$ whose cosine is a rational number, a number $k \in \mathbb{N}$ is called θ-congruent if there is a triangle with rational sides which has θ as an angle and $k\sqrt{s^2 - r^2}$ as its area, where $r \in \mathbb{Z}$ and $s \in \mathbb{N}$ are the unique coprime numbers such that* $\cos(\theta) = r/s$. *Somewhat more precisely we call a triplet $(r, s, k) \in \mathbb{Z} \times \mathbb{N} \times \mathbb{N}$ a solution of the generalized congruent number problem if k is θ-congruent where* $\cos\theta = r/s$ *in lowest terms.*

Scaling the sides of a triangle by a factor a changes the area by the factor a^2; hence a natural number is θ-congruent for some angle θ if and only if its squarefree part is. In other words, (r, s, k) is a solution of the generalized congruent number problem if and only if (r, s, a^2k) is for any number $a \in \mathbb{N}$. Clearly, $(\pi/2)$-congruence is just ordinary congruence of numbers; the only other angles θ for which θ-congruence has been studied somewhat systematically are $\theta = \pi/3$ and $\theta = 2\pi/3$; see [14–17].

The two problems are closely related, as is shown by the following theorem (whose elementary proof we omit).

Theorem 1. *Let* N *be the set of solutions of the generalized congruent number problem, and let* F *be the set of solutions of the concordant form problem. Then mutually inverse bijections* f:N→F *and* g:N→F *are given by*

$$f(r,s,k) := \begin{cases} (s-r, s+r, k) & \text{if } r \not\equiv s \bmod 2, \\ ((s-r)/2, (s+r)/2, 2k) & \text{if } r, s \text{ are both odd} \end{cases}$$

and

$$g(p,q,k) := \begin{cases} ((q+p)/2, (q-p)/2, k) & \text{if } p, q \text{ are both odd}, \\ (q+p, q-p, k/2) & \text{if } p \not\equiv q \bmod 2. \end{cases}$$

We will now proceed to establish a one-to-one correspondence between the intersection of quadrics given by concordant forms and rational points on elliptic curves in standard form. The main difference between our approach and that of other authors such as [3,4,9], (see the next section of this paper) is the use of a true isomorphism (which sets up a one-to-one correspondence between nontrivial solutions of the concordant form problem and points of order greater than 2 on the associated curve) rather than a mapping of degree 4 (which causes the loss of solutions associated with 4-torsion points). The use of a true isomorphism seems more natural and also sheds some new light on the theory of θ-congruent triangles; for example, two rational points of order >2 on the elliptic curve in question give rise to the same θ-congruent triangle if and only if they differ by a 2-torsion element, and a θ-congruent triangle is isosceles if and only if the associated rational point on the curve has order 4 (see Theorem 8 below).

Our approach allows us to clarify two statements in [12] and to answer a question left open in that paper, and we also see that the case $n = 1$ needs to be ruled out in Proposition 5.4 in [11]. (See below for details.)

2. CONNECTIONS TO ELLIPTIC CURVES

In the introduction we exhibited a correspondence between θ-congruent numbers, rational squares in arithmetic progressions and concordant forms written in a certain way. Now we establish a one-to-one correspondence between the intersection of quadrics given by concordant forms and elliptic curves in standard form.

Theorem 2. *Let $m \neq n$ be nonzero integers. We denote by $Q(m, n)$ the set of all $(X_0, X_1, X_2, X_3) \in \mathbb{P}_3(\mathbb{Q})$ such that*

$$X_0^2 + mX_1^2 = X_2^2 \quad \text{and} \quad X_0^2 + nX_1^2 = X_3^2.$$

Also, we denote by $E(m, n)$ the set of all $(T, X, Y) \in \mathbb{P}_2(\mathbb{Q})$ such that

$$Y^2T = X(X + mT)(X + nT),$$

which, in affine notation, is just the elliptic curve $y^2 = x(x + m)(x + n)$. Then mutually inverse isomorphisms $\varphi : Q(m, n) \to E(m, n)$ and $\psi : E(m, n) \to Q(m, n)$ are given by

$$\varphi : \begin{bmatrix} X_0 \\ X_1 \\ X_2 \\ X_3 \end{bmatrix} \mapsto \begin{bmatrix} nX_2 - mX_3 + (m-n)X_0 \\ mn(X_3 - X_2) \\ mn(m-n)X_1 \end{bmatrix}$$

and

$$\psi : \begin{bmatrix} T \\ X \\ Y \end{bmatrix} \mapsto \begin{bmatrix} -(X+mT)\big(Y^2 - m(X+nT)^2\big) \\ 2Y(X+nT)(X+mT) \\ -(X+mT)\big(Y^2 + m(X+nT)^2\big) \\ -(X+nT)\big(Y^2 + n(X+mT)^2\big) \end{bmatrix}$$

Note that φ needs to be redefined at (1, 0, 1, 1) whereas ψ needs to be redefined at (1, $-m$, 0), (1, $-n$, 0) and (0, 0, 1) to obtain well-defined regular maps. Specifically, φ can be rewritten as

$$\varphi(X_0, X_1, X_2, X_3) = \begin{bmatrix} X_1(X_3 - X_2) \\ nX_1(X_0 + X_2) - mX_1(X_0 + X_3) \\ (m - n)(X_0 + X_2)(X_0 + X_3) \end{bmatrix}$$

around (1, 0, 1, 1), whereas ψ can be rewritten as

$$\psi(T, X, Y) = \begin{bmatrix} -(X+mT)\big(X(X+mT) - mT(X+nT)\big) \\ 2YT(X+mT) \\ -(X+mT)\big(X(X+mT) + mT(X+nT)\big) \\ -T\big(Y^2 + n(X+mT)^2\big) \end{bmatrix}$$

which is defined at (1, $-n$, 0) and as

$$\psi(T, X, Y) = \begin{bmatrix} Y^2 - m(X+nT)^2 \\ -2Y(X+nT) \\ Y^2 + m(X+nT)^2 \\ Y^2 - m(X+nT)^2 + 2n(X+mT)(X+nT) \end{bmatrix}$$

which is defined at (0, 0, 1) and at (1, $-m$, 0). We omit the proof that φ and ψ are in fact well-defined and have the desired properties (which is obtained by straightforward calculations) and merely remark that this isomorphism is an instance of a general correspondence between elliptic curves and intersections of quadrics; see [18], p. 36, and [19], pp. 123–125. We note that the isomorphism ψ maps the point at infinity and the 2-torsion points $(-m, 0)$, $(-n, 0)$ and $(0, 0)$ of $E(m, n)$ exactly to the trivial solutions $(1, 0, \pm 1, \pm 1)$ of $Q(m, n)$. Consequently, all other rational points of $E(m, n)$ correspond to nontrivial points on $Q(m, n)$. Thus the following is true (*cf.* [12], Proposition 3; [13], Theorem 1).

Theorem 3. *Let $m \neq n$ be nonzero integers. Then the quadratic forms $X^2 + mY^2$ and $X^2 + nY^2$ are concordant if and only if $E(m, n)$ possesses elements of (finite or infinite) order greater than two.*

We note that the j-invariant of $E(m, n)$ can be easily computed (*cf.* [20], p. 71, or [21], pp. 42–43) to be $j\ E(m, n) = 256(m^2 - mn + n^2)^3/(m^2n^2(m - n)^2)$. The fact that our mappings φ and ψ are true isomorphisms make them more suitable than other correspondences studied in the literature. Let us explain this statement in some detail.

The book [9] deals with the classical congruent number problem. If n is a natural number and (X, Y, Z) are rational sides of a right triangle with area n (where $X < Y < Z$), then the assignment $(X, Y, Z) \mapsto (Z^2/2, (X^2 - Y^2)Z/8) = (x, y)$ (see [9], Chapter I, Paragraph 2, Proposition 2; Chapter I, Paragraph 9, Proposition 19) maps the right triangles to rational points of the elliptic curve $E(-n, n)$ given by the affine equation $y^2 = x^3 - n^2x$. The assignment $(X_0, X_1, X_2, X_3) \mapsto (X_2/X_1, X_0/X_1, X_3/X_1)$ associates with any point (X_0, X_1, X_2, X_3) on the intersection of the quadrics $X_0^2 - nX_1^2 = X_2^2$ and $X_0^2 + nX_1^2 = X_3^2$ with $X_i \geq 0$ and $X_1 > 0$ such a triangle. The composite function τ which is computed to be

$$(X_0, X_1, X_2, X_3) \mapsto \left(\frac{X_0^2}{X_1^2}, -\frac{X_0 X_2 X_3}{X_1^3}\right)$$

obviously extends to a regular morphism $Q(-n, n) \to E(-n, n)$. The trivial elements of $Q(-n, n)$ are mapped to the point at infinity, *i.e.*, to the neutral element of the elliptic curve $E(-n, n)$. This morphism, however, is not an isomorphism, but is a mapping of degree 4 whose image is exactly $2E(-n, n)$, *i.e.*, the set of all doubled points of $E(-n, n)$. If we denote by $\mathbb{D}$ the doubling $P \mapsto 2P$ on the elliptic curve $E(-n, n)$, then an easy computation shows that the following diagram commutes.

$$\begin{array}{ccc} Q(-n,n) & \xrightarrow{\quad\tau\quad} & E(-n,n) \\ & \searrow^{\varphi} \qquad \nearrow_{\mathbb{D}} & \\ & E(-n,n) & \end{array}$$

In other words, for any point S in $Q(-n, n)$ we have $\tau(S) = 2\varphi(S)$. Since the only rational torsion points on $E(-n, n)$ are the points of order 2, these points are mapped to the neutral element of $E(-n, n)$. This diagram explains why, for example, the points $(41^2/7^2, 29520/7^3)$ on $E(-31, 31)$ and $(25/4, 75/8)$ on $E(-5, 5)$ (*cf.* [9], p. 7) are not in the image of τ. However, they correspond to solutions of the associated concordant form problem via the mapping φ. Note that the defect of the mapping τ of not being an isomorphism does not affect any of the assertions in [9]. In particular, the statement that n is a congruent number if and only if $E(-n, n)$ contains rational points of infinite order is true, since the only points of finite order (the 2-torsion points) are mapped to the neutral element of $E(-n, n)$ via the mapping τ, and they correspond to the trivial solutions of the concordant form problem and the degenerate right triangle. However, the defect of τ would make itself felt if one considered not only rational points on the elliptic curve in question but also solutions over number fields.

The use of a correspondence which is not an isomorphism runs into problems when torsion points of order greater than 2 occur, which is the case in the general concordant form problem considered in [3]. The mapping $\sigma : Q(m, n) \to E(m, n)$ used in [3] is given by

$$(X_0, X_1, X_2, X_3) \mapsto \left(\frac{X_0^2}{X_1^2}, \frac{X_0 X_2 X_3}{X_1^3}\right)$$

and hence is the same as the one considered by Koblitz up to the sign in the second component. It turns out that σ is again the composition of the isomorphism φ with an algebraically defined endomorphism of $E(m, n)$, namely the negative $-\mathbb{D}$ of the doubling mapping $\mathbb{D}$. Again, this mapping is of degree 4 and has $2E(m, n)$ as its image. However, in this more general situation, the curve $E(m, n)$ may have points of order 4, and each such point is mapped to a 2-torsion point in the image of σ. Now the 2-torsion points correspond to nontrivial solutions of the associated concordant form problem. Thus, for any pair of numbers (m, n) for which the torsion subgroup of the elliptic curve $E(m, n)$ is isomorphic to $\mathbb{Z}_2 \times \mathbb{Z}_4$, there exist nontrivial solutions of the concordant form problem in the sense of Definition 1.1 for the pair (m, n) even if the rank of $E(m, n)$ is zero. Hence qualification is needed for the statement in [3], which specifies that if $E(m, n)$ has rank zero then nontrivial solutions exist if and only if the torsion group is $\mathbb{Z}_2 \times \mathbb{Z}_8$ or $\mathbb{Z}_2 \times \mathbb{Z}_6$ ([3], p. 101). For example, if $(m, n) = (-1, 3)$ then $(X, Y, Z, W) = (\pm 1, \pm 1, 0, \pm 2)$ are solutions of the equations $X^2 + mY^2 = Z^2$ and $X^2 + nY^2 = W^2$ corresponding to the 4-torsion points $(3, \pm 6)$ and $(-1, \pm 2)$ of the elliptic curve $y^2 = x(x - 1)(x + 3)$; note that $E(-1, 3)$ has rank zero (see [14]). These solutions are not covered by Main Corollary 1 (pp.

104–105) and Corollary 2 (p. 107) in [3] where only solutions are considered for which all components are nonzero.

We note in passing that the curve $y^2 = x(x - 1)(x + 3)$ also provides a counterexample to Proposition 5.4 in [11], which does not hold for $n = 1$. In fact, $n = 1$ is a $\pi/3$-congruent number (realized by the equilateral triangle with all sides equal to 2), which, in fact, corresponds to the 4-torsion points $(3, \pm 6)$ and $(-1, \pm 2)$ of the elliptic curve $E(-1, 3)$. Hence the additional assumption $n > 1$ is indispensible to make Proposition 5.4 in [11] correct.

In the recent paper [4], again only solutions with nonzero components are considered, as becomes clear from Definition 2 in that paper (which should presumably state that a solution (X, Y, Z, W) is considered nontrivial only if $XY\,ZW \neq 0$). With this definition of nontriviality the assertions in [4] (especially Corollary 12 on p. 797) are correct since no torsion points of orders 3, 6 and 8 occur in the cases studied in [4]. However, 4-torsion points are possible, which correspond to solutions of the associated system of quadrics in which one component is zero, but which differ from the truly trivial solutions $(\pm 1, 0, \pm 1, \pm 1)$. For example, for $m = 1$ and $n = k^2$ where $k \in \{2, 3, 4, 5, 6, 8, 9, 13\}$ solutions exist (namely $(X, Y, Z, W) = (0, 1, 1, k)$), which are not covered in [4]. The condition that the rank of $E(1, k^2)$ be zero is satisfied for the quoted values of k, whereas the rank is one for $k \in \{7, 10, 11, 12\}$, as we verified with the Sage Mathematical Software System (© 2005–2012 William Stein and the Sage Development Team; http://www.sagemath.org). It seems unfortunate to us to rule out such solutions, as they correspond to objects in other contexts (rational θ-triangles, rational squares in arithmetic progressions) with interesting properties (*cf.* Theorem 8 below).

While a thorough analysis of the mapping σ used in [3] together with the two-descent on the elliptic curve $E(m, n)$ may reveal all the interesting phenomena concerning solutions to the concordant form problem from the study of rational points on $E(m, n)$ via σ, the approach via the isomorphism φ seems to be much more direct and natural. Also, the correspondence between solutions (r, s, k) of the congruent number problem and rational points of order >2 on the elliptic curve $y^2 = x(x - (s - r)k)(x + (s + r)k)$ becomes, via the correspondence with rational squares in arithmetic progressions, more lucid than in [12–14]. It also becomes clear from our calculations that two rational points of order > 2 on the curve yield the same triangle if and only if they differ by a 2-torsion element in the Mordell–Weil group of the curve. Moreover, our calculations clarify Theorem 1 and Proposition 4 in [12]. More precisely, we will exactly determine those numbers n occurring as θ-congruent numbers corresponding to torsion elements of the associated elliptic curve, which was left open in [12] and was clarified in the paper [22], of which we were made aware only after finishing our paper.

3. NONTRIVIAL TORSION SOLUTIONS

When we speak of the torsion subgroup of an elliptic curve E over $\mathbb{Q}$, we always mean the torsion subgroup of the Mordell–Weil group $E_{\mathbb{Q}}$ of rational points on E. A deep theorem by Mazur (see [23]) states that the torsion subgroup of any elliptic curve over $\mathbb{Q}$ must be one of the groups $\mathbb{Z}_m$ where $1 \le m \le 10$ or $m = 12$ or else of the groups $\mathbb{Z}_2 \times \mathbb{Z}_{2n}$ where $1 \le n \le 4$. Since $E(m, n)$ has three points of order 2, namely $(0, 0)$, $(-m, 0)$ and $(-n, 0)$, only the last four possibilities can occur for the curves studied here. We want to determine the exact conditions on m and n which determine the type of the torsion group. To do so, we compute all nontrivial torsion elements. This was essentially already done in [3], and only a few additional calculations (omitted here) are needed to arrive at the following complete characterization of all torsion elements. (Also see [22].)

Theorem 4. *We consider the elliptic curve* $E(m,\ n)$ *over* $\mathbb{Q}$ *where* $m = -pk$ *and* $n = qk$ *such that* $p,\ q \in \mathbb{N}$ *are coprime and* $k \in \mathbb{N}$ *is squarefree.*

- *There are points of order* 4 *if and only if* $-m$ *and* $n-m$ *are squares, say* $-m = u^2$ *and* $n = v^2 - u^2$*. In this case, the* 4*-torsion points are exactly the four points*

$$\left(u^2 - uv,\ \pm v(u^2 - uv)\right) \quad \textit{and} \quad \left(u^2 + uv,\ \pm v(u^2 + uv)\right).$$

- *There are points of order* 8 *if and only if there are numbers* $\xi,\ \eta \in \mathbb{N}$ *such that* $\xi^2 + \eta^2$ *is a square, say* $\xi^2 + \eta^2 = \zeta^2$*, and the equations* $m = -\xi^4$ *and* $n = \eta^4 - \xi^4 = \zeta^2(\eta^2 - \xi^2)$ *hold. In this case, the* 8*-torsion points are exactly the eight points*

$$\begin{aligned}
&(\ \xi\zeta(\xi+\eta)(\zeta+\eta),\ \pm\xi\eta\zeta(\xi+\eta)(\zeta+\xi)(\zeta+\eta)\ ),\\
&(\ \xi\zeta(\xi+\eta)(\zeta-\eta),\ \pm\xi\eta\zeta(\xi+\eta)(\zeta-\xi)(\zeta-\eta)\ ),\\
&(\ \xi\zeta(\xi-\eta)(\zeta+\eta),\ \pm\xi\eta\zeta(\xi-\eta)(\zeta-\xi)(\zeta+\eta)\ ),\\
&(\ \xi\zeta(\xi-\eta)(\zeta-\eta),\ \pm\xi\eta\zeta(\xi-\eta)(\zeta-\eta)(\zeta+\xi)\ ).
\end{aligned}$$

- *There are points of order* 3 *(or, equivalently, points of order* 6*) if and only if there are coprime integers* $a,\ b \neq 0$ *with* $a + 2b \neq 0$, $b + 2a \neq 0$ *and* $a \pm b \neq 0$ *such that* $m = a^3(a + 2b)$ *and* $n = b^3(b + 2a)$*. In this case the points of order* 3 *are the two points*

$$(a^2b^2,\ \pm a^2b^2(a+b)^2),$$

and the points of order 6 *are the six points*

$$\left(-a^2b(b+2a),\ \pm a^2b(b+2a)(a^2-b^2) \right),$$
$$\left(-ab^2(a+2b),\ \pm ab^2(a+2b)(a^2-b^2) \right),$$
$$\left(ab(a+2b)(b+2a),\ \pm ab(a+2b)(b+2a)(a+b)^2 \right).$$

In all other cases for m and n the only torsion points are the trivial points (0, 0), (−*m*, 0) *and* (−*n*, 0).

Remark 1. *If* $k = d^2\ell$ *is not squarefree, then the elliptic curves* $E(-p\ell, q\ell)$ *and* $E(-pk, qk)$ *are isomorphic (as algebraic groups) via the isomorphism* $(x, y) \mapsto (d^2x, d^3y)$. *Thus, the torsion subgroups of the corresponding Mordell–Weil groups of rational points are also isomorphic. As a special case, we note that the curve* $E(-pk, qk)$ *contains points of order* 4 *if and only if* $-m = pk$ *and* $n - m = qk + pk$ *are squares (irrespectively of whether or not k is squarefree).*

We are now ready to give a complete classification of all concordant forms and θ-congruent triangles that correspond to torsion solutions of the associated elliptic curve. This classification is based on the following theorem (*cf.* [22]).

Theorem 5. *We consider the elliptic curve* $E(m, n)$ *where* $m = -pk$ *and* $n = qk$ *such that* $p, q \in \mathbb{N}$ *are coprime and* $k \in \mathbb{N}$ *is squarefree. Let T be the torsion subgroup of* $E(m, n)$.

i. *If* $T \cong \mathbb{Z}_2 \times \mathbb{Z}_8$ *or* $T \cong \mathbb{Z}_2 \times \mathbb{Z}_8$ *then* $k = 1$.

ii. *If* $T \cong \mathbb{Z}_2 \times \mathbb{Z}_6$ *then* $k = 1$ *or* $k = 3$.

(*The examples in the next section will show that each of the possible cases occurs for an infinite number of elliptic curves* $E(m, n)$).

Proof. Let us first consider the case that that $T \cong \mathbb{Z}_2 \times \mathbb{Z}_4$ or $T \cong \mathbb{Z} \times \mathbb{Z}_8$. Assume $k \neq 1$; then there is a prime divisor t of k. The number t then divides both $-m = pk$ and $n - m = (p + q)k$, which are squares according to Theorem 4 (i) and (ii). Since k is squarefree, this implies that t divides both p and $p + q$, which is impossible because p and q are coprime by assumption. Thus the assumption $k \neq 1$ is wrong, and we must have $k = 1$ in this case.

Let us now consider the case that $T \cong \mathbb{Z}_2 \times \mathbb{Z}_6$. By Theorem 4 (iii) there are coprime numbers $a, b \in \mathbb{Z} \setminus \{0\}$ with $a + 2b \neq 0$, $b + 2a \neq 0$ and $a \pm b \neq 0$ such that $m = a^4 + 2a^3b$ and $n = 2ab^3 + b^4$. Let t be a prime divisor of k. Then t divides both $-pk = m = a^3(a + 2b)$ and $qk = n = b^3(2a + b)$. It is obvious that if the prime t were a divisor of a then it would also be a divisor of b, and vice versa; this, however, is impossible because a and b are coprime. Thus t divides neither a nor b, hence divides both $a + 2b$ and $2a + b$, hence divides $2(a + 2b) = (2a + b) + 3b$, hence divides $3b$ and consequently must be 3. We have shown that 3 is the only possible prime divisor of k. This implies that $k = 1$ or $k = 3$. □

The implication of this theorem for concordant forms and θ-congruent numbers will be elucidated in the next section.

4. INTERPRETATION AND CONCLUSIONS

Theorem 5 may be interpreted in terms of the concordant form problem and of the generalized congruent number problem.

Theorem 6. *Let $m = -pk$ and $n = qk$ such that $p, q \in \mathbb{N}$ are coprime and $k \in \mathbb{N}$ is squarefree.*

i. *If the quadratic forms* X^2+mY^2 *and* X^2+nY^2 *are concordant due to 4- or 8-torsion, then* $k = 1$.

ii. *If the quadratic forms* X^2+mY^2 *and* X^2+nY^2 *are concordant due to 3- or 6-torsion then* $k = 1$ *or* $k = 3$.

Proof. This is an immediate consequence of Theorem 5. □

Theorem 7. *Let* $\theta = \arccos(r/s)$ *where* $r \in \mathbb{Z}$ *and* $s \in \mathbb{N}$ *are coprime numbers such that* $|r| < s$. *Moreover, let* $k \in \mathbb{N}$ *be a squarefree number.*

- *If k is odd and is a* θ*-congruent number due to 4- or 8-torsion then* $k = 1$.
- *If k is even and is a* θ*-congruent number due to 4- or 8-torsion then* $k = 2$.
- *If k is odd and is a* θ*-congruent number due to 3- or 6-torsion then* $k = 1$ *or* $k = 3$.
- *If k is even and is a* θ*-congruent number due to 3- or 6-torsion then* $k = 2$ *or* $k = 6$.

Proof. We remember that the associated elliptic curve is given by $E(m, n)$ where $m = -(s - r)k$ and $n = (s + r)k$ if $r \not\equiv s \bmod 2$ (first case) and where $m = -(s - r)k/2$ and $n = (s + r)k/2$ if r, s are both odd (second case). In the first case, the coefficients satisfy the hypotheses of Theorem 5 and we see that in the situation of (i) we have $k = 1$ and for (iii) we have $k = 1$ or $k = 3$. The same holds in the second case when k is odd. It remains to consider the second case with $k = 2\ell$ being even (and ℓ being odd since $k$is assumed to be squarefree). In this situation the elliptic curve $E(m', n')$ associated with the triplet (r, s, ℓ) is given by $m' = -(s - r)\ell/2$ and $n' = (s + r)\ell/2$, and these coefficients satisfy the hypotheses of

Theorem 5. So we can conclude that $\ell = 1$ (and hence$k = 2$) in the situation (ii) and $\ell = 1$ or $\ell = 3$ (and hence $k = 2$ or $k = 6$) in the situation (iv). □

Any of the above situations occurs for infinitely many elliptic curves. The characterization of the curves with a prescribed torsion subgroup, together with the arguments in the proof of Theorem 5, gives rise to several series of examples which will illustrate this fact.

Example 1. Torsion solutions of order 4. Let u, v be any coprime numbers with $u < v$ and let $m = -u^2$ and $n = v^2-u^2$. Then the torsion subgroup $T \subseteq E_{\mathbb{Q}}(m, n)$ contains $\mathbb{Z}_2 \times \mathbb{Z}_4$. Hence any point of order 4 defines a solution of the concordant form problem given by the triplet $(p, q, k) = (-m, n, 1)$. If m and n are both odd (which is equivalent to saying that v is even) then this point defines a solution to the generalized congruent number problem given by the triplet $(r, s, k) = ((n-m)/2, (n+m)/2, 1)$ as well. If m and n have different parities, then the elliptic curve $E(4m, 4n)$ also has a point of order 4, which defines a solution to the generalized congruent number problem given by the triplet $(n - m, n + m, 2)$. Explicit instances are given by $(u, v) = (1, 2)$, which corresponds to $(m, n) = (-1, 3)$ and represents the situation that v is even, and by $(u, v) = (1, 3)$, which corresponds to $(m, n) = (-1, 8)$ and represents the situation that v is odd.

Example 2. Torsion solutions of order 8. Let (ξ, η, ζ) be any primitive Pythagorean triplet so that $\xi^2 + \eta^2 = \zeta^2$ and $\xi < \eta$ (note that ξ and η then automatically have unequal parities!) and let $m = -\xi^4$ and $n = \eta^4 - \xi^4$. Then the torsion subgroup $T \subseteq E_{\mathbb{Q}}(m, n)$ is isomorphic to $\mathbb{Z}_2 \times \mathbb{Z}_8$. Thus, any point of order 8 defines a solution of the concordant form problem given by the triplet $(p, q, k) = (-m, n, 1)$. If m and n are both odd (which means ξ is odd and η is even), then this point also defines a solution of the generalized congruent number problem given by $(r, s, k) = ((n-m)/2, (n+m)/2, 1)$. If m is even (and a fortiori n is odd), then the elliptic curve E(4m, 4n) also has points of order 8, each of which defines a solution of the generalized congruent number problem given by the triplet (n − m, n + m, 2). Explicit instances are given by $(\xi, \eta, \zeta) = (3, 4, 5)$, which corresponds to (m, n) = (−81, 175) and represents the situation of a primitive Pythagorean triplet with ξ being odd, and by by $(\xi, \eta, \zeta) = (8, 15, 17)$, which corresponds to (m, n) = (−4096, 46 529) and represents the situation of a primitive Pythagorean triplet with ξ being even.

Example 3. Torsion solutions of order 3 or 6. *Let a, b be any coprime numbers with* $a < 0$, $b > 0$, $a + 2b > 0$, $2a + b > 0$ *and* $a + b \neq 0$, *and let* $m = a^3(a + 2b)$ *and* $n = b^3(2a + b)$. *Then the torsion subgroup* $T \subseteq E_{\mathbb{Q}}(m, n)$ *is isomorphic to* $\mathbb{Z}_2 \times \mathbb{Z}_6$. *From the proof of Theorem 5 we know that the only possible common divisors of m and n are 1 or 3.*

- First case: gcd(a + 2b, 2a + b) = 1. In this case any point of order 3 or 6 defines a solution of the concordant form problem given by the triplet (p, q, k) = (−m, n, 1). If m and n are both odd then this point also defines a solution of the generalized congruent number problem given by (r, s, k) = ((n − m)/2, (n + m)/2, 1). If $m \not\equiv n \bmod 2$ then the elliptic curve E(4m, 4n) also has points of order 3 and 6, which define solutions of the generalized congruent number problem given by the triplet (n − m, n + m, 2).

- Second case: gcd(a + 2b, 2a + b) = 3. (Note that this situation occurs when $a \equiv b \bmod 3$.) Let p = −m/3 and q = n/3. Then any point of order 3 or 6 defines a solution of the concordant form problem given by the triplet (p, q, k) = (−m/3, n/3, 3). If m and n are both odd, then this point also defines a solution of the generalized congruent number problem given by (r, s, k) = ((n − m)/6, (n + m)/6, 3). If $m \not\equiv n \bmod 2$ then the elliptic curve E(4m, 4n) also has points of order 3 and 6, which define solutions to the generalized congruent number problem given by the triplet $((n - m)/3, (n + m)/3, 6)$.

- Explicit instances are given as follows:
- the example $(a, b) = (-1, 3)$ corresponds to $(m, n) = (-5, 27)$ and represents the situation that a, b are odd coprime numbers with different signs satisfying the congruence condition $a \not\equiv b \mod 3$ and the inequalities $a + 2b > 0$ and $2a + b > 0$;
- the example $(a, b) = (-2, 5)$ corresponds to $(m, n) = (-64, 125)$ and represents the situation that a, b are coprime numbers with different signs and different parities satisfying the congruence condition $a \not\equiv b \mod 3$ and the inequalities $a + 2b > 0$ and $2a + b > 0$;
- the example $(a, b) = (-5, 13)$ corresponds to $(m, n) = (-875 \cdot 3, 2197 \cdot 3)$ and represents the situation that a, b are odd coprime numbers with different signs satisfying the congruence condition $a \equiv b \mod 3$ and the inequalities $a + 2b > 0$ and $2a + b > 0$;
- the example $(a, b) = (-2, 7)$ corresponds to $(m, n) = (-32 \cdot 3, 343 \cdot 3)$ and represents the situation that a, b are coprime numbers with different signs and different parities satisfying the congruence condition $a \equiv b \mod 3$ and the inequalities $a + 2b > 0$ and $2a + b > 0$.

Remark 2. From the above considerations it is clear that each of the possible situations occurs for an infinite number of cases. Moreover, for any of these cases it is easy to construct series of examples such that the j-invariants of the corresponding elliptic curves are pairwise different.

Furthermore, with these examples we answer a question left open in [12] (Remark 1 after Proposition 4). Namely, in [12] it is shown that if n ∉ {1, 2, 3, 6} then n is a θ-congruent number if and only if the rank of the associated elliptic curve is positive; i.e., it is not possible to obtain a corresponding θ-triangle from torsion points on this elliptic curve. In [12] it was shown that the condition n ∉ {1, 2}is indispensible in this statement; i.e., there are torsion solutions for n = 1 and n = 2. For n = 3 and n = 6 this was left as an open problem, which is now answered affirmatively by the above considerations.

The following result yields a nice geometric characterization of the 4-torsion solutions.

Theorem 8. Let p, q ∈ ℕ be coprime, let k ∈ ℕ be squarefree, and let m = −pk and n = qk. Consider a rational point P on the curve E(m, n), the associated θ-congruent triangle Δ and the associated triplet T of rational squares in an arithmetic progression. Then the following statements are equivalent:

i. *P has order four;*

ii. *T contains the number zero;*

iii. *Δ is isosceles, i.e., has two equal sides.*

Proof. Under the correspondences in Theorem 1, we have $a = b$ for the sides of Δ if and only if $\alpha = 0$ for the smallest element in the progression $\alpha^2 < \beta^2 < \gamma^2$, which is the case if and only if $X_2 = 0$ for a corresponding point (X_0, X_1, X_2, X_3) on$Q(m, n)$ as defined in Theorem 2 (since $\alpha = X_2/X_1$, $\beta = X_0/X_1$ and $\gamma = X_3/X_1$ under the correspondence between points on this curve and rational squares in arithmetic progression). Now the condition $X_2 = 0$ corresponds to the equation $(x + m)(y^2 + m(x + n)^2) = 0$ via the biregular mapping in Theorem 2. This equation yields either $x = -m$ (which is one of the 2-torsion points) or $y^2 + m(x+n)^2 = 0.$. In the latter case $-m$ must clearly be a square. Since both $y^2 = -m(x+n)^2$ and $y^2 = x(x+m)(x+n)$, we have

$$\begin{aligned} 0 &= x(x+m)(x+n) + m(x+n)^2 \\ &= (x+n)(x^2 + 2mx + mn) \\ &= (x+n)\big((x+m)^2 + m(n-m)\big) \end{aligned}$$

and hence $(x+m)^2 = -m(n-m).$. Thus, not only $-m$ but also $n - m$ must be a square. This is exactly the condition that $E(m, n)$ contains points with 4-torsion. Writing $-m = u^2$, the equation $y^2 = -m(x+n)^2 = u^2(x+n)^2$ yields $y = \pm u(x + n)$, which shows that the points (x, y) on $E(m, n)$ associated with isosceles triangles are exactly the points of order 4. □

The isosceles triangles occurring as solutions of the generalized congruent number problem can be characterized in terms of intrinsic geometrical properties, as we now show. (Note that when we speak of an isosceles θ-triangle we always assume the angle θ to be between the two equal sides.)

Theorem 9. *There is an isosceles rational θ-triangle if and only if* $\sin(\theta/2)$ *is rational. If* $\sin(\theta/2) = \varrho/\sigma$ *where %, $\sigma \in \mathbb{N}$ are coprime, then the sides of the unique rational θ-triangle with squarefree k are given by a, a, c with $a = k\sigma$ and $c = 2ak$* $\sin(\theta/2) = 2k\varrho$*.*

Proof. First, if a, a, c are the rational sides of an isosceles rational θ-triangle with a rational cosine $\cos(\theta) = r/s$, then $c = 2a\sin(\theta/2)$ so that $\sin(\theta/2)$ is also rational. Conversely, if $\sin(\theta/2)$ is rational, say $\sin(\theta/2) = \varrho/\sigma$ with $\varrho, \sigma \in \mathbb{N}$ coprime, then $\cos(\theta) = 1 - 2\sin^2(\theta/2)$ is rational. If σ is odd, then $\cos(\theta) = (\sigma^2 - 2\varrho^2)/\sigma^2 = r/s$ is a coprime representation with r, s both odd.

Let $a = 2\sigma$ and $c = 2a\sin(\theta/2) = 4\varrho$. Then the isosceles rational θ-triangle with sides a, a, c has the area $(a^2/2)\sin(\theta) = 2\sigma^2\sqrt{1-\cos^2(\theta)} = 2\sigma^2\sqrt{(s^2-r^2)/s^2} = (2\sigma^2/s)\sqrt{s^2-r^2} = 2\sqrt{s^2-r^2}$, hence this triangle is the unique solution with $k = 2$.

Remark 3. The equilateral triangle with all three sides of length 2 plays a somewhat special role. First, it is the unique isosceles rational π/3-triangle with area$k\sqrt{3}$ and squarefree $k \in \mathbb{N}$; in fact, $k = 1$. Furthermore, there is no other rational π/3-triangle with area$\sqrt{3}$, since the rank of the Mordell–Weil group $E_{\mathbb{Q}}(-1, 3)$ is zero (cf. [3,14]). For any squarefree $k > 1$ there are either no rational π/3-triangles with area$k\sqrt{3}$ at all, or else there are infinitely many such triangles. The only case in which there is a single solution is the case $k = 1$ with the abovementioned equilateral triangle.

The possibility of moving freely between the concordant form problem and the generalized congruent number problem provides a way of translating solutions found for one of these problems to solutions of the other problem. For example, various interesting examples of concordant forms and of triplets of rational squares in arithmetic progressions can be obtained from the examples for rational $2\pi/3$-triangles found in [13]. Conversely, examples for concordant forms given in [3] can be used to construct rational θ-triangles.

REFERENCES

1. Euler, L. De binis formulis speciei xx + myy et xx + nyy inter se concordibus et discordibus. *Mem. Acad. Sci. St.-Pet.*1780, 8, pp. 3–16. Available online: https://math.dartmouth.edu/~euler/docs/originals/E758.pdf accessed on January 16, 2015.

2. Ono, T. *Variations on a Theme of Euler*; Plenum Press: New York, NY, USA, 1994.

3. Ono, K. Euler's Concordant Forms. *Acta Arith.* **1996**, *78*, 101–123.

4. Im, B.-H. Concordant Numbers within Arithmetic Progressions and Elliptic Curves. *Proc. Amer. Math. Soc.* **2013**, *141*, 791–800.

5. Conrad, K. *Arithmetic Progression of Four Squares*, Available online: http://www.math.uconn.edu/kconrad/ross2007/4squarearithprog.pdf accessed on 12 January 2015.

6. Gonzàles-Jiménez, E.; Steuding, J. Arithmetic progressions of four squares over quadratic fields. *Publ. Math. Debrecen.* **2010**, *77*, 125–138.

7. Gonzàles-Jiménez, E.; Xarles, X. Five Squares in Arithmetic Progression over Quadratic Fields. *ArXiv E-Prints* **2009**,*arXiv*, 0909.1663.

8. Xarles, X. Squares in Arithmetic Progression over Number Fields. *J. Number Theory* **2012**, *132*, 379–389.

9. Koblitz, N. *Introduction to Elliptic Curves and Modular Forms*; Springer: Berlin, Germany, 1993.

10. Tunnell, J.B. A classical Diophantine problem and modular forms of weight 3/2. *Invent. Math.* **1983**, *72*, 323–334.

11. Top, J.; Yui, N. Congruent number problems and their variants. *Algorithmic Number Theory: Lattices, Number Fields, Curves and Cryptography*, Math. *Sci. Res. Inst. Publ.* **2008**, *44*, 613–639.

12. Fujiwara, M. θ-congruent numbers. In *Number Theory*; Győry, K., Halasz, G., Eds.; Walter de Gruyter: Berlin, Germany, 1998; pp. 235–241.

13. Kan, M. θ-congruent numbers and elliptic curves. *Acta Arith.* **2000**, *94*, 153–160.

14. Yoshida, S. Some Variants of the Congruent Number Problem II. *Kyushu J. Math.* **2001**, *55*, 387–404.

15. Yoshida, S. Variants of the Congruent Number Problem III. *Kyushu J. Math.* **2002**, *56*, 147–165.

16. Yoshida, S. Some Variants of the Congruent Number Problem III. In *Technical Reports of Mathematical Sciences*; Chiba University: Chiba, Japan, 2008; Volume 24.

17. Janfada, A.S.; Salami, S.; Dujella, A.; Peral, J.C. On the high rank $\pi/3$- and $2\pi/3$-congruent number elliptic curves.*ArXiv E-Prints.* **2011**, *arXiv*, 1102.4291.

18. Cassels, J.W.S. *Lectures on Elliptic Curves*; Cambridge University Press: Cambridge, UK, 1991.

19. Connell, I. *Elliptic Curve Handbook*, Available online: http://www.math.mcgill.ca/connell/ accessed on 12 January 2015.

20. Husemöller, D. *Elliptic Curves*; Springer: New York, NY, USA, 2004.

21. Silverman, J.H. *The Arithmetic of Elliptic Curves*, 2nd ed; Springer: Dordrecht, Netherlands, 2009.

22. Fujiwara, M. Some properties of θ-congruent numbers. In *Natural Science Report*; Ochanomizu University: Tokyo, Japan; February; 2002; Volume 52, 2, pp. 1–8.

23. Mazur, B. Modular curves and the Eisenstein ideal. *Publ. Math. I.H.E.S* **1977**, *47*, 33–186.

CHAPTER 11

A Study on the Nourishing Number of Graphs and Graph Powers

Sudev Naduvath [1,*] and Germina Augustine [2]

[1]Department of Mathematics, Vidya Academy of Science & Technology, Thalakkottukara, Thrissur 680501, Kerala, India

[2]PG & Research Department of Mathematics, Mary Matha Arts & Science College, Mananthavady, Wayanad 670645, Kerala, India

ABSTRACT

Let $\mathbb{N}_0$ be the set of all non-negative integers and P(N0) be its power set. Then, an integer additive set-indexer (IASI) of a given graph G is defined as an injective function f:V(G)→P(N0) such that the induced edge-function f+:E(G)→P(N0) defined by $f^+(uv) = f(u) + f(v)$ is also injective, where $f(u) + f(v)$ is the sumset of $f(u)$

and $f(v)$. An IASI f of G is said to be a strong IASI of G if $|f^+(uv)| = |f(u)|\,|f(v)|$ for all $uv \in E(G)$. The nourishing number of a graph G is the minimum order of the maximal complete subgraph of G so that G admits a strong IASI. In this paper, we study the characteristics of certain graph classes and graph powers that admit strong integer additive set-indexers and determine their corresponding nourishing numbers.

KEYWORDS

graph powers; integer additive set-indexers; strong integer additive set-indexers; nourishing number of a graph

1. INTRODUCTION

For all terms and definitions, not defined specifically in this paper, we refer to [1–4]. For more about different graph classes, we further refer to [5–7]. Unless mentioned otherwise, all graphs considered here are simple, finite and have no isolated vertices.

Let $\mathbb{N}_0$ denote the set of all non-negative integers. For all $A, B \subseteq \mathbb{N}_0$, define their sumset, denoted by $A + B$, as the set $A + B = \{a + b : a \in A, b \in B\}$. If either A or B is countably infinite, then their sumset is also countably infinite. Hence, the sets we consider here are all finite sets of non-negative integers. The cardinality of a set A is denoted by $|A|$. The following theorem is on the minimum and maximum of the cardinality of the sumset of two non-empty sets of integers.

Theorem 1. [8] *For any two non-empty sets A and B of integers,* $|A| + |B| - 1 \leq |A + B| \leq |A|\,|B|$.

Based on the concepts of sumsets of finite non-empty sets, an integer additive set-indexer of a given graph G is defined as follows.

Definition 1. [9] An *integer additive set-indexer* (IASI, in short) of a given graph G is an injective function $f:V(G)\to P(N0)$ such that the induced edge-function $f+:E(G)\to P(N0)$ defined by $f^+(uv) = f(u) + f(v)$ is also injective. A graph which admits an IASI is called an integer additive set-indexed graph (strong IASI-graph).

The *set-indexing number* (see [10]) of an element (a vertex or an edge) of G is the cardinality of the set-label of that element. The IASI f of a given graph G such that the set-indexing number of every edge of G is the product of the set-indexing numbers of its end vertices is of special interest and hence we have defined the following.

Definition 2. [11] If a graph G has a set-indexer $f:V(G)\to P(N0)$ such that $|f^+(uv)| = |f(u) + f(v)| = |f(u)|\,|f(v)|$ for all adjacent vertices u and v of G, then f is said to be a *strong IASI* of G. A graph which admits a strong IASI is called a *strong integer additive set-indexed graph* (IASI-graph).

We use the notation $A < B$ in the sense that $A \cap B = \emptyset$. We notice that the relation $<$ is symmetric, but not reflexive and need not be transitive. By the sequence $A_1 < A_2 < A_3 < \ldots < A_n$, we mean that the given sets are pairwise disjoint.

The *difference set* of a set A, denoted by D_A is the set of all numerical differences between any two elements of A. That is $D_A = \{|a - b| : a, b \in A\}$. The following lemma establishes a necessary and sufficient condition for the cardinality of the sumset $A + B$ of two non-empty sets A and B of integers to be the product of the cardinalities of A and B.

Lemma 1. [11] *Let A, B be two non-empty subsets of* $\mathbb{N}_0$. *Then,* $|A + B| = |A|\,|B|$ *if and only if their difference sets, denoted by* D_A *and* D_B *respectively, follow the relation* $D_A < D_B$ (*That is,* $D_A \cap D_B = \emptyset$).

Theorem 2. [11] *Let each vertex* v_i *of the complete graph* K_n *be labeled by the set* $A_i \in P(N0)$. *Then* K_n *admits a strong IASI if and only if there exists a finite sequence of sets* $D_1 < D_2 < D_3 < \dots < D_n$ *where each* D_i *is the difference set of the set-label* A_i *of the vertex* v_i.

In other words, a complete graph K_n admits a strong IASI if and only if the difference sets of the set-labels of all vertices of K_n are pairwise disjoint. Hence, as a subgraph of a complete graph, any given graph G admit a strong IASI and hence we have the following theorem.

Theorem 3. [11] *A connected graph G (on n vertices) admits a strong IASI if and only if each vertex* v_i *of G is labeled by a set* A_i *in* $P(N0)$ *and there exists a finite sequence of sets* $D_1 < D_2 < D_3 < \dots < D_m$, *where* $m \le n$ *is a positive integer and each* D_i *is the set of all differences between any two elements of the set* A_i.

The relation < of difference sets defined above is called the *difference relation* on G. The finite sequence $D_1 < D_2 < D_3 < \dots < D_m$ of difference sets of the set-labels of vertices of a given graph G is called *difference chain* of G.

Theorem 4. [11] *If a graph G admits a strong IASI, then any subgraph* G_1 *of G also admits strong IASI, which is an IASI induced on* G_1 *by G.*

In view of the above notions, we have introduced the following notion.

Definition 3. [12] The *nourishing number* of a set-labeled graph is the minimum length of the maximal chain of difference sets in G. The nourishing number of a graph G is denoted by $\varkappa(G)$. That is, the nourishing number of a given graph G is minimum order of the maximal subset of G, where the difference relation, of the difference sets of all of whose vertices is an equivalence relation.

Theorem 5. [12] *The difference sets of the set-labels of all vertices of a complete graph K_n are pairwise disjoint. That is, the nourishing number of a complete graph K_n is n.*

Remark 1. Let G_1 be a clique in a strong IASI graph G. Then by Theorem 4, G_1 also admits the induced strong IASI. Then, the set-labels of all vertices of G_1 have pairwise disjoint difference sets. Therefore, the nourishing number of a graph G is the minimum order of a maximal complete subgraph of G.

Invoking the above remark, the following results have been established.

Theorem 6. [12], *The nourishing number of any triangle-free graphs is* 2.

Hence, we have

Remark 2. *In view of Theorem 6, the nourishing number of trees, cycles and bipartite graphs is* 2.

A characterisation of strong IASI graphs was done in [11–13]. As an extension to these studies, in this paper, we discuss the nourishing number of certain graph classes and the graph powers.

2. NEW RESULTS

We consider the r-th power of a graph only if r is a positive integer. Now, recall the definition of graph powers.

Definition 4. [1] The r-th power of a simple graph G is the graph G^r whose vertex set is V, two distinct vertices being adjacent in G^r if and only if their distance in G is at most r. The graph G^2 is referred to as the *square* of G, the graph G^3 as the*cube* of G.

The following theorem on graph powers is the most relevant result in our present discussion.

Theorem 7. [14] *If d is the diameter of a graph G, then G^d is a complete graph.*

In view of Theorem 7, we proceed to determine the nourishing number of different graph classes and their powers. The following remark provides the nourishing number of complete graphs and their different powers.

Remark 3. Any power of a complete graph K_n is K_n itself and hence by Theorem 5, the nourishing number of any power ofK_n is n itself.

Hence, we proceed with the study about the nourishing number of a finite powers of complete bipartite graphs.

Proposition 1. *The nourishing number of the r-th power of a complete bipartite graph is*

$$\varkappa(K^r_{m,n}) = \begin{cases} 2 & \text{if } r = 1 \\ m+n & \text{if } r \geq 2. \end{cases}$$

Proof. By Remark 2, the nourishing number of a complete bipartite graph G is 2. The diameter of a complete graph G is 2. Therefore, by Theorem 7, G^2 is a complete graph on $m + n$ vertices. Hence, for an IASI defined on the graph G^r and hence the difference sets of the set-labels of all vertices of G^r must be pairwise disjoint. Therefore, ϰ(Krm,n)=m+n, for $r \geq 2$. □

Now, we study the nourishing number of paths P_m on m vertices. By Theorem 6, the nourishing number of paths is 2. Hence, we need to study the nourishing number of arbitrary powers of paths. The following result determines the nourishing number of the r-th power of a path.

Theorem 8. *The nourishing number of the r-th power of a path P_m is*

$$\varkappa(P_m) = \begin{cases} r+1 & \textit{if } r < m \\ m+1 & \textit{if } r \geq m. \end{cases}$$

Proof. Let $P_m = v_1v_2v_3 \ldots v_n$; $n = m + 1$. The diameter of P_m is m. Then by Theorem 7, Pmm is a complete graph on $m + 1$ vertices. Therefore, by Theorem 5, for $r \geq m$, the nourishing number of the r-th power of P_m is $m + 1$.

If $r < m$, then for $1 \leq i \leq n$, the vertices in the set $V' = \{v_i, v_{i+1}, v_{i+2}, \ldots, v_{i+r}\}$ are pairwise adjacent and the hence the induced subgraph $\langle V' \rangle$ of Prm is a complete graph on $r + 1$ vertices. More over, since the vertices v_i and v_{i+r+1} are not adjacent in Prm, there exists no complete subgraph of order $r + 1$ or higher in Prm. Hence the difference sets of the set-labels of all vertices in V' are pairwise disjoint. Since i is arbitrary, we have the order of a maximal complete subgraph in Prm is r+1. Therefore, difference sets of at least $r + 1$ set-labels of mutually adjacent vertices of G must be pairwise disjoint. Hence, by Theorem 5, the nourishing number of Prm is $r + 1$. □

We now proceed to discuss nourishing number of finite powers of cycle graphs. By Theorem 6, the nourishing numbers of cycles are also 2. Hence, we need to study about the nourishing number of certain higher integral powers of cycles. The following result discusses the nourishing number of the r-th power of a cycle C_n on n vertices.

Theorem 9. *The nourishing number of the r-th power of a cycle* C_n *is*

$$\varkappa(C_n) = \begin{cases} r+1 & \text{if } r < \lfloor \frac{n}{2} \rfloor \\ n & \text{if } r \geq \lfloor \frac{n}{2} \rfloor. \end{cases}$$

Proof. Let $C_n = v_1v_2v_3 \ldots\ v_nv_1$. The diameter of C_n is ⌊n2⌋. Then by Theorem 7, C⌊n2⌋n is a complete graph on n vertices. Therefore, by Theorem 5, for r≥⌊n2⌋, the nourishing number of the r-th power of C_n is n.

If r≥⌊n2⌋, then for $1 \leq i \leq n$, the vertices in the set $V' = \{v_i, v_{i+1}, v_{i+2}, \ldots, v_{i+r}\}$ are pairwise adjacent in Crn and the hence the induced subgraph $\langle V' \rangle$ of Crn is a complete graph on $r + 1$ vertices. Also, since the vertices v_i and v_{i+r+1} are not adjacent in Crn, there is no complete subgraph of higher order induced by the set of vertices $\{v_i, v_{i+1}, v_{i+2} \ldots, v_{i+r+1}\}$. Since the choice of $i$is arbitrary, we have the size of a maximal complete subgraph in Crn is $r + 1$. Hence, by Theorem 5, the difference sets of the set-labels of all vertices in V' are pairwise disjoint and hence the nourishing number of Crn is $r +$ 1, for any positive integer r≥⌊n2⌋. □

A graph that arises interest in this context is a wheel graph which is defined as follows.

Definition 5. [3] A *wheel graph*, denoted by W_{n+1}, is defined as $W_{n+1} = C_n + K_1$.

The nourishing number of finite powers of wheel graphs is determined in the following proposition.

Proposition 2. *The nourishing number of the r-th power of a wheel graph is*

$$\varkappa(W^r_{n+1}) = \begin{cases} 3 & \textit{if } r = 1 \\ n+1 & \textit{if } r \geq 2. \end{cases}$$

Proof. We can determine the nourishing number of W_{n+1} by taking $r = 1$. Let $K_1 = v$ and $C_n = v_1v_2v_3 \ldots v_nv_1$. For $1 \leq i \leq n$, three vertices v, v_i and v_{i+1} form a triangle in W_{n+1}. Since v_i and v_{i+2} are not adjacent in W_{n+1}, this triangle is the maximal clique in W_{n+1}, where the set-labels of these three vertices have disjoint difference sets. Therefore, $\varkappa(W_{n+1}) = 3$.

Since any two vertices of a wheel graph W_{n+1} are at a distance 2 from each other, the square of it is a complete graph on $n+1$ vertices. Hence, by Theorem 5, the difference sets of the set-labels of all vertices of G^2 are pairwise disjoint. Therefore, ϰ(Wrn+1)=n+1 for all $r \geq 2$. □

Another interesting graph which is related to a wheel graph is a *helm graph* which is defined as follows.

Definition 6. [6] A *helm graph*, denoted by H_n, is a graph obtained by adjoining a pendant edge to each of the vertices of the outer cycle of a wheel graph W_{n+1}. A helm graph has $2n+1$ vertices and $3n$ edges.

The following theorem discusses the nourishing number of a helm graph and its powers.

Theorem 10. *The nourishing number of the r-th power of a helm graph H_n is given by*

$$\varkappa H_n^r = \begin{cases} 3 & if\ \ r = 1 \\ n+1 & if\ \ r = 2 \\ n+4 & if\ \ r = 3 \\ 2n+1 & if\ \ r \geq 4. \end{cases}$$

Proof. Let u be the central vertex, $V = \{v_1, v_2, v_3,..., v_n\}$ be the set of vertices of the cycle C_n and $W = \{w_1, w_2, w_3,..., w_n\}$ be the set of pendant vertices in H_n. In H_n, the central vertex u is adjacent to each vertex v_i of V and each v_i is adjacent to a vertex w_i in W.

The maximal clique in H_n is a triangle formed by the vertices $vv_iv_{i+1}v$, for $1 \leq i \leq n$ (in the sense that $v_{n+1} = v_1$), the difference sets of whose set-labels are pairwise disjoint. Therefore, $\varkappa(H_n) = 3$.

Since each vertex v_i in V is adjacent to u, the distance between any two vertices in H_n is 2. Hence, the subgraph of H2ninduced by $V \cup \{u\}$ is a complete graph on n + 1 vertices in H2n. Since the distances between any two w_is is greater than 2, there is no complete graph of higher order in H2n. Therefore, ϰ(H2n)=n+1.

Since each vertex w_i in W is at a distance at most 3 from u as well as from all vertices of V, for $1 \leq i \leq n$, the subgraph of H3ninduced by $V \cup \{u, w_{i-1}, w_i, w_{i+1}\}$ is a complete graph. Since the distances between all other vertices in W and W_i are greater than 3, there is no complete graph of higher order. Therefore, $\varkappa(H_n^3) = n + 4.$

Since u is adjacent all $v_i \in V$, the distance between any two vertices in H_n is 4. Therefore, $H_n^r;\ r \geq 4$ is a complete graph on $2n$ + 1 vertices. Hence by Theorem 5, for $r \geq 4$, the nourishing number of Hrn is $2n$ + 1. □

Another graph we consider in this context is a friendship graph whose definition is give below.

Definition 7. [6,7] A *friendship graph* F_n is the graph obtained by joining n copies of the cycle C_3 with a common vertex. It has $2n + 1$ vertices and $3n$ edges.

The following proposition is about the nourishing number of a friendship graph F_n.

Proposition 3. *The nourishing number of the r-th power of a friendship graph F_n is*

$$\varkappa(F_n^r) = \begin{cases} 3 & \text{if } r = 1 \\ 2n+1 & \text{if } r \geq 2 \end{cases}$$

Proof. The maximal clique in a friendship graph F_n is a triangle. Then, by Theorem 5, $\varkappa(F_n) = 3$. Since all the vertices in $F_n$at a distance 1 or 2 among themselves, by Theorem 7, F2n is a complete graph on $2n + 1$ vertices. Hence, by Theorem 5, $\varkappa(F_n^r) = 2n + 1, \text{if } r \geq 2.$

Another similar graph structure is a fan graph which is defined as follows.

Definition 8. [5,6] A *fan graph*, denoted by $F_{m,n}$, is defined as the graph join $\bar{K}_m + P_n.$

Proposition 4. *The nourishing number of the r-th power of a fan graph $F_{m,n}$ is*

$$\varkappa(F_{m,n}^r) = \begin{cases} 3 & \text{if } r = 1 \\ m+n & \text{if } r \geq 2 \end{cases}$$

Proof. The maximal clique in a fan graph $F_{m,n}$ is a triangle. Then, by Theorem 5, $\varkappa(F_n) = 3$. Since all the vertices in $F_{m,n}$ at a distance 1 or 2 among themselves and hence by Theorem 7, $F^2_{m,n}$ is a complete graph on $m + n$ vertices. Hence, by Theorem 5, $\varkappa(F^r_{m,n}) = m + n$ for all $r \geq 2$. □

Now, we proceed to discuss the nourishing number of split graphs and and their powers. A *split graph* is defined as follows.

Definition 9. [5] A *split graph*, denoted by $G(K_r, S)$, is a graph in which the vertices can be partitioned into a clique K_r and an independent set S. A split graph is said to be a *complete split graph* if every vertex of the independent set S is adjacent to every vertex of the the clique K_r and is denoted by $K_S(r, s)$, where r and s are the orders of K_r and S respectively.

The following theorem estimates the nourishing number of a split graph.

Theorem 11. *The nourishing number of the r-th power of a split graph $G = G(K_r, S)^r$ is given by*

$$\varkappa(G^r) = \begin{cases} r & \text{if no vertex of } S \text{ is adjacent to all vertices of } K_r \\ r+1 & \text{if some vertices of } S \text{ are adjacent to all vertices of } K_r \\ r+l & \text{if } r = 2 \\ r+s & \text{if } r \geq 3; \end{cases}$$

where l is the maximum number of vertices of S which are adjacent to the same vertex of K_r and s is the number of non-isolated vertices in S.

Proof. Let G be a split graph consisting of a clique K_r and an independent set S that contains s vertices. The isolated vertices in S do not have any impact on the adjacency in the powers of G and hence we need not consider isolated vertices in S. Hence, without loss of generality, let S has no isolated vertices in G.

If there is no vertex of S which has adjacency with every vertex of K_r, then K_r itself is the maximal clique in G and hence $\varkappa(G) = r$.

Assume that some vertices of S are adjacent to all vertices of K_r. Let w_i be a vertex in S that is adjacent to every vertex of K_r. Then $K_r + \{w_i\}$ is a complete subgraph of G. Since no vertices in S are adjacent among themselves, This subgraph is a maximal clique in G. Therefore, $\varkappa(G) = r + 1$.

Let $S_1 = \{w_1, w_2, \ldots, w_l\}$ be the set of vertices in S that are adjacent to a vertex, say v_i in K_r. Then each of them is at distance 2 among themselves and from all other vertices of K_r. Therefore, $G_1 = K_r \cup S_1$ is a complete graph in G^2. Since, S_1 is the maximal set of that kind, G_1 is a maximal clique in G^2. Therefore, $\varkappa(G^2) = r + l$.

All vertices in a split graph G are at a distance at most 3 in G. Therefore, G^3 is a complete graph on $r + s$ vertices. Therefore, $\varkappa(G^r) = r + s$, for all values of $r \geq 3$. □

In view of Theorem 11, we establish the following result on complete split graphs.

Corollary 1. *The nourishing number of the r-th power of a complete split graph $G = K_S(r, s)$ is given by*

$$\varkappa(G^r) = \begin{cases} r+1 & \text{if } k = 1 \\ r+s & \text{if } k \geq 2; \end{cases}$$

Proof. Since every vertex of S is adjacent to all vertices of K_r in G, by Theorem 11, $\varkappa(G) = r + 1$. Also, in G, being adjacent to all vertices of K_r, all vertices in S are at a distance 2 among themselves, $K_r \cup S$ is a complete graph on $r + s$ vertices in G^2. Therefore, $\varkappa(G^k) = r + s$ for all $k \geq 2$. □

Another important graph class we need to consider in this context is a sun graph which is defined as follows.

Definition 10. [5,7] An *n-sun* or a *trampoline*, denoted by S_n, is a chordal graph on $2n$ vertices, where $n \geq 3$, whose vertex set can be partitioned into two sets $U = \{u_1, u_2, u_3, \dots, u_n\}$ and $W = \{w_1, w_2, w_3, \dots, w_n\}$ such that W is an independent set of $G$and w_j is adjacent to u_i if and only if $j = i$ or $j = i + 1 \pmod n$. *A complete sun* is a sun G where the induced subgraph $\langle U \rangle$ is complete.

The sun graphs with $\langle U \rangle$ is a cycle is one of the most interesting structures among the sun graphs. The following theorem determines the nourishing number of a sun graph G whose non-independent set of vertices induces a cycle in the graph G.

Theorem 12. *If G is an n-sun graph with $\langle U \rangle = C_n$, then the nourishing number of G^r is given by*

$$\varkappa(G^r) = \begin{cases} 2r+1 & \text{if } r < \lfloor \frac{n}{2} \rfloor \\ 2(n-1) & \text{if } r = \lfloor \frac{n}{2} \rfloor \text{ and if } n \text{ is odd} \\ 2n-1 & \text{if } r = \lfloor \frac{n}{2} \rfloor \text{ and if } n \text{ is even} \\ 2n & \text{if } r \geq \lfloor \frac{n}{2} \rfloor + 1; \end{cases}$$

Proof. Let $r < \lfloor \frac{n}{2} \rfloor$, then the subgraph of G^r induced by the vertex set $V' = \{w_{n-r+1}, w_{n-r+2}, w_{n-r+3}, \dots, w_n, v_{n-r+1}, v_{n-r+2}, v_{n-r+3}, \dots, v_n, v_1\}$ is a clique of order $2r + 1$ in G^r. Since neither w_{n-r} is adjacent to v_{i+1} nor w_{n-r+1} is adjacent to v_{i+2}, this clique is a maximal clique in G^r. Therefore, by Theorem 5, $\varkappa(G^r) = 2r + 1$.

If $r = \lfloor \frac{n}{2} \rfloor$ and n is odd, then the vertex w_n is adjacent to every vertex of V and to the vertices of W except w_r and w_{r+1}.

Then, $V' = V \cup [W - \{w_r, w_{r+1}\}]$ induces a maximal clique of order $2n - 2$ in G. Therefore, by Theorem 5, if $r=\lfloor n2 \rfloor$ and n is odd, then the nourishing number of G^r is $2(n - 1)$.

If $r=\lfloor n2 \rfloor+1$ and n is even, then the vertex w_n is adjacent to every vertex of V and to the vertices of W except w_r in G^r. Then, $V' = V \cup [W - \{w_r\}]$ induces a maximal clique of order $2n - 1$ in G^r. Therefore by Theorem 5, if $r=\left\lfloor \frac{n}{2} \right\rfloor$ and n is odd, then the nourishing number of G^r is $2n - 1$.

Since all the vertices in $G = S_n$ are at a distance at most $r=\left\lfloor \frac{n}{2} \right\rfloor$, G^r is a complete graph on $2n$ vertices, where $r \geq 1+\left\lfloor \frac{n}{2} \right\rfloor$. Hence, by Theorem 5, for $r \geq \left\lfloor \frac{n}{2} \right\rfloor+1$, the nourishing number of G^r is $2n$. □

The following theorem discusses the nourishing number of a complete n-sun graph.

Theorem 13. *The nourishing number of a complete n-sun graph G is given by*

$$\varkappa(G^r) = \begin{cases} n & \text{if } r = 1 \\ n+1 & \text{if } r = 2 \\ 2n & \text{if } r \geq 3; \end{cases}$$

Proof. If $r = 1$, K_n itself is the largest clique in the the complete sun graph G. Therefore, $\varkappa(G) = n$.

Each vertex w_i in the independent set W is adjacent to two vertices of U and is at a distance 2 from all other vertices of U. Hence, each vertex w_i is adjacent to all vertices of U in G^r. The subgraph of G^2 induced by $U \cup \{w_i\}$ is a clique in G^2. Since no two vertices in W are at a distance 2, this clique is a maximal clique in G^2. Therefore, $\varkappa(G^2) = n + 1$.

Since all vertices in G are at a distance at most 3, the cube and higher powers of G are complete graphs on $2n$ vertices. Therefore, by 5, $\varkappa(G^r) = 2n$ for $r \geq 3$. □

Another important graph involving cycle graphs is a sunlet graph. An n-sunlet graph is defined as follows.

Definition 11. [7] The *n-sunlet* graph is the graph on $2n$ vertices obtained by attaching one pendant edge to each vertex of a cycle C_n.

The following result discusses about the nourishing number of an n-sunlet graph.

Theorem 14. *The nourishing number of the r-th power of an n-sunlet graph G is*

$$\varkappa(G^r) = \begin{cases} 2r & \text{if } r < \lfloor \frac{n}{2} \rfloor + 1 \\ 2(n-1) & \text{if } r = \lfloor \frac{n}{2} \rfloor + 1 \text{ and } n \text{ is odd} \\ 2n-1 & \text{if } r = \lfloor \frac{n}{2} \rfloor + 1 \text{ and } n \text{ is even} \\ 2n & \text{if } r \geq \lfloor \frac{n}{2} \rfloor + 2. \end{cases}$$

Proof. The proof of this proposition is similar to the proof of Theorem 12. Let $V = \{v_1, v_2, v_3, \ldots, v_n\}$ be the vertex set of the cycle C_n and $W = \{w_1, w_2, w_3, \ldots w_n\}$ be the set of pendant vertices in G.

If $r < \lfloor \frac{n}{2} \rfloor + 1$, the vertex set $V' = \{w_{n-r+2}, w_{n-r+3}, w_{n-r+4} \ldots, w_n, v_{n-r+1}, v_{n-r+2}, v_{n-r+3}, \ldots, v_n, v_1\}$ induces a clique in G on $2r$ vertices. Since there is no other vertex outside V' which is at a distance at most r from all vertices in V', the induced graph $\langle V' \rangle$ is a maximal clique in G^r. Therefore, for $r<1+\lfloor \frac{n}{2} \rfloor$, $\varkappa(G) = 2r$.

If $r=\lfloor \frac{n}{2} \rfloor+1$ and n is odd, then the vertex w_n is adjacent to every vertex of V and to the vertices of W except w_r and w_{r+1}. Then, $V' = V \cup [W - \{w_r, w_{r+1}\}]$ induces a maximal clique of order $2n - 2$ in G. Therefore, in this case, the nourishing number of G^r is $2(n - 1)$.

If $r=\lfloor \frac{n}{2} \rfloor+1$ and n is even, then the vertex w_n is adjacent to every vertex of V and to the vertices of W except w_r in G^r. Then, $V' = V \cup [W - \{w_r\}]$ induces a maximal clique of order $2n - 1$ in G^r. Therefore, in this case, the nourishing number of G^r is $2n - 1$.

Since the maximum distance between the vertices in W is $2+\lfloor \frac{n}{2} \rfloor$, diameter of G is $2+\lfloor \frac{n}{2} \rfloor$. Hence, by Theorem 7, G^r is a complete graph for all $r \geq 2+\lfloor n2 \rfloor$ on $2n$ vertices. Therefore, the difference sets of the set-labels of all vertices of G^r are pairwise disjoint. Hence, for $r \geq 2+\lfloor \frac{n}{2} \rfloor$ the nourishing number of G^r is $2n$. □

3. CONCLUSIONS

In this paper, we have established some results on the nourishing number of certain graphs and graph powers. The admissibility of strong IASI by various other graph classes, graph operations and graph products and finding the corresponding nourishing numbers are open. Determining the nourishing numbers of powers of bipartite graphs, chordal graphs, crown graphs, prism graphs etc. are some of the open problems.

Besides the above-mentioned problems, there are several other graph classes and structures for which the admissibility of strong

IASIs are yet to be verified and the corresponding nourishing numbers are to be determined. More over, the properties and characteristics of strong IASIs, both uniform and non-uniform, are yet to be investigated. The problems of establishing the necessary and sufficient conditions for various graphs and graph classes to have certain IASIs are also open.

ACKNOWLEDGMENTS

The authors would like to thank the anonymous reviewers for their insightful suggestions and critical and constructive remarks which made the overall presentation of this paper better.

AUTHOR CONTRIBUTIONS

Both authors contributed significantly in writing this article. Both authors read and approved the final manuscript.

CONFLICTS OF INTEREST

The authors declare no conflict of interest regarding the publication of this paper.

REFERENCES

1. Bondy, J.A.; Murty, U.S.R. *Graph Theory*; Springer: Berlin, Germany, 2008.

2. Chartrand, G.; Zhang, P. *Introduction to Graph Theory*; McGraw-Hill Inc: New York, NY, USA, 2005.

3. Harary, H. *Graph Theory*; Addison-Wesley Publishing Company Inc.: Boston, MA, USA, 1994.

4. West, D.B. *Introduction to Graph Theory*; Pearson Education Inc.: Upper Saddle River, NJ, USA, 2001. [

5. Brandstädt, A.; Le, V.B.; Spinard, J.P. *Graph Classes: A Survey*; SIAM: Philadelphia, PA, USA, 1999.

6. Gallian, J.A. A Dynamic Survey of Graph Labeling. *Electron. J. Comb.* **2013**, #DS6.

7. *Information System on Graph Classes and Their Inclusions*, Available online: http://www.graphclasses.org accessed on 22 February 2015.

8. Nathanson, M.B. *Additive Number Theory: Inverse Problems & Geometry of Sumsets*; Springer: Berlin, Germany, 1996.

9. Germina, K.A.; Anandavally, T.M.K. Integer Additive Set-Indexers of a Graph: Sum Square Graphs. *J. Comb. Inf. Syst. Sci.* **2012**, *37*, 345–358.

10. Germina, K.A.; Sudev, N.K. On Weakly Uniform Integer Additive Set-Indexers of Graphs. *Int. Math. Forum.* **2013**, *8*, 1827–1834.

11. Sudev, N.K.; Germina, K.A. Some New Results on Strong Integer Additive Set-Indexers of Graphs. *Discret. Math. Algorithms Appl.* **2015**, *7*.

12. Sudev, N.K.; Germina, K.A. A Characterisation of Strong Integer Additive Set-Indexers of Graphs. *Commun. Math. Appl.* **2014**, *5*, 91–100.

13. Sudev, N.K.; Germina, K.A. On Integer Additive Set-Indexers of Graphs. *Int. J. Math. Sci. Eng. Appl.* **2014**, *8*, 11–22.

Index

T